SMALL-SCALE PAPER-MAKING

Small-scale paper-making

Practical
ACTION
PUBLISHING

Practical Action Publishing Ltd
25 Albert Street, Rugby,
Warwickshire, CV21 2SD, UK
www.practicalactionpublishing.com

First published in 1985
First published by Practical Action Publishing in 1993
Transferred to digital printing in 2008

A catalogue record for this book is available from the British Library & Library of Congress

ISBN 978-1-85339-189-7 Paperback
ISBN 978-1-78044-594-6 Digital book

Citation: 1985 *Small-scale Papermaking*, Rugby, UK: Practical Action Publishing
https://doi.org/10.3362/9781780445946

Since 1974, Practical Action Publishing has published and disseminated books and information
in support of international development work throughout the world. All print editions are
produced and distributed via ethical and sustainable print on demand global facilities.

Practical Action Publishing is a trading name of Practical Action Publishing Ltd (Company Reg.
No. 01159018 | VAT 880 9924 76). All profits are covenanted back to its parent group, Practical
Action (Charity Reg. No. 247257).

The manufacturer's authorised representative in the EU for product safety is Lightning Source
France, 1 Av. Johannes Gutenberg, 78310 Maurepas, France. compliance@lightningsource.fr

PREFACE

The satisfaction of society's needs for printed information and cultural publications requires an important expansion of the current paper production capacity of most developing countries. While an adequate level of per capita consumption is estimated at approximately 30 kg per year of various types of paper, current per capita consumption in a large number of developing countries does not exceed 2 to 5 kg per year. If the desirable level is to be attained by the end of this century and if imports are to be simultaneously reduced, paper production will therefore have to increase at quite a fast rate.

Imports of some types of paper may however remain unavoidable for small and medium-sized developing countries in view of their limited demand within the country by comparison with the very large capacity of existing paper mills. This is particularly the case of newsprint, the production of which benefits from economies of scale and usually requires large timber resources. There is, however, a wide range of writing, printing and packaging paper which could be produced within the country instead of being imported. Public planners will then need to decide whether production of these types of paper should take place in large plants using imported technology, equipment and expertise, or in a number of small paper mills which offer various advantages in terms of employment generation, rural industrialisation or increased use of locally available materials. A third possibility for sufficiently large countries or groups of countries might be to establish both large and small plants producing different types and qualities of paper products. For example, this course has been adopted in India where small and large paper mills cater for different segments of the market. The purpose of this memorandum is to provide technical and economic information on these various possibilities with a view to assisting persons directly or indirectly interested in paper production in making appropriate choices.

Chapter I provides information on the effects of small-scale and large-scale paper production on employment generation, foreign exchange expenditures, backward and forward linkages, rural industrialisation, the environment and so on. This chapter, which will be of particular interest to public planners, concludes with some guidelines for government action in favour of appropriate paper-making technologies and scales of production. The chapter will also be of interest to practising or would-be paper producers since it provides information on capital costs, foreign exchange inputs, labour requirements and the efficiency of alternative paper-making technologies.

Chapters II to V provide a conspectus of paper-making operations, mostly in small mills which are defined for the purposes of this memorandum as mills with a capacity not exceeding 30 tonnes of paper per day. The raw materials used by these mills include agricultural residues, waste paper, waste cotton and rags, and imported wood-based pulps. Besides the characteristics of various grades of paper and raw materials, these chapters describe a variety of paper-making technologies and equipment in some detail. However, this memorandum does not constitute a technical introduction to paper-making in general or a basic textbook on the subject; it provides criteria of choice of methods for people already in possession of the necessary background knowledge. The information provided in this volume, unlike that to be found in technical memoranda on simpler subjects prepared by the ILO and UNIDO, is not sufficient for the establishment or operation of the small-scale production facilities suggested. Established or would-be paper manufacturers need to obtain additional information and assistance from equipment suppliers, engineering firms or specialised institutions. For this purpose, the names and addresses of some of them are provided in separate appendices.

Chapter VI provides a methodological framework for the evaluation of the profitability of paper mills. The chapter will be of particular interest to paper manufacturers, financial institutions and project evaluators in industrial development agencies.

Comments and observations on the content and usefulness of this publication can conveniently be sent to the ILO or UNIDO by replying to the questionnaire reproduced at the end of the memorandum. They will be taken into consideration in the preparation of additional technical memoranda.

This memorandum was prepared by A. Western (consultant for Intermediate Technology Consultants Ltd., a subsidiary of Intermediate Technology Development Group Limited) and M. Allal, staff member in charge of the series of technical memoranda within the Technology and Employment Branch of the ILO.

A.S. Bhalla,
Chief,
Technology and Employment Branch

.

CONTENTS

A C K N O W L E D G E M E N T S

The publication of this technical memorandum has been made possible by a grant from the Swedish International Development Authority (SIDA). The International Labour Office and the United Nations Industrial Development Organisation gratefully acknowledge this generous support.

CHAPTER I

SOCIO-ECONOMIC EFFECTS OF ALTERNATIVE PAPER
MANUFACTURING TECHNOLOGIES: THE CASE FOR MINI PAPER MILLS

I. INTRODUCTION

Paper consumption may be considered as one measure, among others, of the economic development of a country. Statistics show that there is a very strong correlation between per capita income and per capita consumption of all types of paper products. Per capita consumption in most developing countries is lower than 10 kg/year (e.g. 2.7 kg/year for India and 0.4 kg/year for Ethiopia in 1980) while that of industrialised countries is closer, on average, to 200 kg/year (e.g. 276.2 kg/year for the United States and 191.3 kg/year for Sweden in 1980).

Demand for paper is a function of various variables, including the price of paper, educational expenditures, local customs and so on. However, as stated earlier, per capita income is the most important determinant of paper consumption. This is particularly the case in developing countries where income elasticities of demand for paper are much higher than in industrialised countries. Thus, for the same growth of per capita income, the increase in paper consumption should be much higher in developing countries than in developed countries.

Paper consumption in developing countries will still remain much lower than in industrialised countries for the foreseeable future. However, it is bound to grow at a fairly fast rate if one of a country's development objectives is to achieve full literacy and to satisfy the society's needs for printed information and cultural publications. It is estimated (UNIDO, 1979) that an adequate level of per capita consumption for the fulfilment of the above objective is 30 kg/year. Some developing countries have already reached this level whilst a large number of countries, especially in Africa and Asia, must increase their current consumption tenfold to fiftyfold in order to reach this consumption level. How can this be achieved? Public planners may choose among various alternatives: to increase imports of paper, to produce it in one or a few large-scale plants, to establish a number of small paper mills or to opt for a mixed solution which incorporates the above alternatives. To reach a decision, the government will need to take into consideration a number of factors, including market demand for paper (in terms of both quantity and quality), production costs, foreign exchange savings, employment generation and rural industrialisation. These factors are analysed in the following sections of this chapter in relation to various scales of production, taking into consideration the Indian experience. Reliance on the latter for an assessment of alternative paper-making technologies is justified by the fact that India is one of the very few developing countries, if not the only one, which has made a concerted effort to promote the establishment of small paper mills often called "mini paper plants". Over 100 such plants are now operating in India alongside the large paper mills. Currently, small paper mills produce nearly 20 per cent of the total volume of paper manufactured in India. Their share in the production of paper other than newsprint is expected to reach 40 per cent by 1990. Indian technologists and enterprises have fully mastered the technology used in small paper mills and local engineering firms are well equipped to manufacture the equipment required. Some of the equipment used in large plants is also manufactured in India. Obviously, the Indian experience in paper-making may not be applied to all developing countries without some adaptations. This is particularly true for small countries with limited engineering facilities and an insufficient supply of the type of raw materials used by small mills (e.g. straw, bagasse, waste paper). Nevertheless, the Indian approach provides useful guidelines for public planners who wish to expand paper production through the promotion of paper-making technologies consonant with national development objectives.

The following section of this chapter deals with demand, production and trade in paper since they ultimately determine the scale of production. Section III provides a definition of small scale in paper manufacture. The following sections (IV and V) analyse the effects of alternative paper-making technologies on a number of variables of particular interest to public planners and provide some guidelines for government action in favour of suitable technologies and scales of production.

II. DEMAND, PRODUCTION AND TRADE[1]

Per capita consumption of paper by developing countries in 1979 was estimated at 2 kg/year for Africa, 4.5 kg/year for Asia and 25 kg/year for Latin America. Within each region, national differences in per capita consumption are fairly large, depending on the level of development of each country. For example, per capita consumption in Africa varies between 1 kg/year and 5 kg/year while in Latin America it varies between 1 kg/year and 65 kg/year. Furthermore, per capita consumption in urban areas is generally much higher than in rural areas (e.g. per capita consumption in the large urban areas of India is close to 30 kg/year while the national average is 2.7 kg/year).

Asia and Latin America produce a large fraction of the paper consumed in these two regions (76 per cent for Asia and 82 per cent for Latin America), while over 50 per cent of the paper consumed in Africa is imported. Very few developing countries are net exporters of paper products.

Newsprint, linerboard, and kraftliner represent a relatively large share of total paper imports, ranging from 29 per cent for Africa to 73 per cent for Latin America. This may be explained by the fact that the production of newsprint benefits from economies of scale.

In recent years, the production of paper and paperboard by developing countries has been growing at a rate approximately equal to 7.7 per cent. This rate will need to increase a great deal if most developing countries are to reach the minimum per capita paper consumption of 30 kg/year. It is therefore doubtful whether the large number of developing countries with a current per capita consumption of 5 kg/year or less can reach this minimum level before the end of this century. To achieve this objective, they will need to attain annual growth rates exceeding 10 to 12 per cent.

---

[1] The statistics in this section have been taken from the July 1980 issue of Pulp and Paper International (see Bibliography).

The choice of scale of production is directly affected by the current state of paper production and consumption by developing countries. Let us examine the case of an average developing country with a population of 5 million and a current per capita paper consumption of 4 kg/year. If additional imports are to be avoided, a 10 per cent yearly increase in consumption will call for a production increase of 2,000 tonnes a year. This can be achieved by a small plant with a capacity of 6 to 7 tonnes per day (TPD) or a medium-sized plant (50 TPD or more) operating at a fraction of its capacity for a relatively large number of years. From a purely economic viewpoint, the choice of the latter alternative may be difficult to justify. It will be shown below that low plant capacity utilisation is one factor, among others, which may militate against the establishment of medium-sized or large paper mills in developing countries.

III. DEFINITION OF SMALL-SCALE IN PAPER MANUFACTURE

Since this technical memorandum focuses on small-scale paper manufacturing it is important to define the term "small-scale" as it relates to the paper industry.

A definition of "small-scale" may be based on a review of the capacity of established paper mills, or the paper needs of a country or on the classification used in the national legislation. New mills in North America and Scandinavia are often designed to produce 1,000 tonnes of pulp and paper per day. In comparison, mills with a capacity of 100 TPD or less may be considered as small. On the other hand, for 35 developing countries consuming less than 5,000 tonnes per year of all grades of paper, a 15 TPD mill may be considered fairly large since it can supply all the paper needs of the country. India defines a small paper mill as one with a capacity equal or inferior to 30 TPD. This definition was also adopted at the UNIDO International Forum on Appropriate Industrial Technology held in New Delhi in November 1978. According to this Forum, integrated pulp and paper mills should be classified as follows:

- small : up to 30 TPD
- medium-sized: between 30 and 100 TPD
- large : above 100 TPD

The foregoing definition of "small-scale" is adopted for the purpose of this memorandum, especially since it draws a great deal from the Indian experience.

From a purely technological viewpoint, the upper limit of a small-scale mill could be 50 TPD because the low cost technology used in mills of this capacity cannot be easily applied to much larger mills. Furthermore, mills with a capacity closer to 50 TPD may reduce unit production costs through chemical recovery and/or partial power generation. These additional processes are not economically feasible for mills with capacity lower than 30 TPD. Thus, an upper limit of 50 TPD may be adopted by developing countries for the definition of small-scale mills and the formulation of legislation in favour of small-scale paper-making.

IV. SOCIO-ECONOMIC EFFECTS OF ALTERNATIVE PAPER-MAKING TECHNOLOGIES

The choice of scale of production and of an appropriate paper manufacturing technology is a fairly difficult exercise since both scale and technology affect a large number of socio-economic variables of particular importance to developing countries. The purpose of this section is to analyse the various effects of alternative scales of production and technologies with a view to helping public planners formulate appropriate measures and policies for the local paper industry. This section should also be of interest to financing institutions and to practising or would-be paper producers since it includes estimates of investment and foreign exchange costs, labour inputs and unit production costs.

It may be noted that some of the factors analysed below apply to either integrated pulp and paper mills or to mills which produce paper only with pulp obtained elsewhere. Thus, these factors may not be of interest to some paper producers.

IV.1 Raw materials

The scale of production of a pulp mill will depend, to a large extent, on the type of raw materials which are available. Medium-sized and large mills may not operate profitably if the raw materials consist of agricultural residues (e.g. wheat or rice straw, bagasse), various grasses, waste paper or rags because it is doubtful, for a number of reasons, whether these mills can be supplied with large quantities of the above materials on a regular basis.

First, agricultural residues or bagasse are partly used for other purposes such as animal feed, fertilisers and energy generation. Thus, a pulp mill may only get the surplus agricultural residues unless a relatively high price is paid to farmers or sugar mill owners. Secondly, agricultural residues are available during harvesting seasons only and a mill will require a very large storage capacity to avoid discontinuity of production. Thirdly, the amount of waste paper available for recycling (usually estimated at 20 to 25 per cent of paper consumed in a given area) will never be sufficient for the requirements of a large mill. Finally, an extension of the collection area in order to ensure a regular supply of agricultural residues or waste paper will considerably increase the transport cost of these materials and decrease the economic viability of the mill. Furthermore, the transport infrastructure in many developing countries may not facilitate the transport of raw materials over long distances. For all these reasons, the scale of production of pulp mills which use agricultural residues or waste paper as raw materials must be kept relatively small if they are to be operated at full capacity and if production costs are to remain within acceptable limits.

While small pulp mills will generally be preferred for the processing of agricultural residues or waste paper, the production of wood-based pulps is best carried out in large plants for the following reasons. Firstly, logs must be transformed into wood chips if disc refiners and/or digesters are to be used in the pulping process. Currently, the smallest commercial chipper has a recommended capacity of approximately 15 tonnes per hour. Thus, if this equipment is to be fully used, the pulp mill will have to process over 150 tonnes of wood chips per day (i.e. a quantity far above the capacity of a small pulp mill). Secondly, the use of stone grinders in mechanical pulping requires the processing of whole logs. The capacity of commercial stone grinders also exceeds that of small pulp mills. In sum, the production of wood pulp will generally not be profitable if carried out in small pulp mills.

If species of trees appropriate for the paper industry do not grow in a country, the latter may conceivably import wood in the form of logs or chips if it wishes to establish a large pulping plant. However, it is very doubtful whether such a plant can be operated profitably in view of the high transport cost of the raw material. A country without appropriate timber resources should therefore either import pulp or process agricultural residues and waste paper in small pulp mills. Alternatively, countries with appropriate climatic conditions could grow the right species of trees for the paper industry. This alternative should be carefully investigated in view of the very high investment costs required, including infrastructural costs for countries without adequate transport facilities.

Developing countries with appropriate timber resources could establish large pulping plants. Indeed, a number of such countries in Latin America and Asia have adopted this solution. Other countries which may contemplate establishing similar plants should investigate the following:

- whether it is equally or more profitable to use trees for the production of pulp than for other purposes, such as the production of various types of wood products; and
- whether the full plant capacity does not exceed the local demand for pulp, and in the affirmative, whether the excess pulp can be exported at competitive prices.

In addition to availability, the choice of raw materials for the paper industry should take into consideration additional benefits which may be derived from the use of agricultural residues or waste paper by small pulp mills. These may include additional employment generated by the collection and transport of these materials, additional revenues to farmers who do not have alternative outlets for surplus agricultural residues, and improvement of the profitability of local sugar mills which may sell their excess bagasse to pulp mills. It should however be added that the use of agricultural waste by small pulp mills may also present some disadvantages, including the irregular supply of sufficient volumes of these wastes, the need for large storage areas, and the possibility of high price increases in the case of bad harvests.

A small pulp mill need not use a single source of raw materials. Straw, bagasse and waste paper may, for example, be used simultaneously by the same mill in addition to imported wood pulp. The final composition of the pulp will depend on such factors as the local availability of various raw materials, the type and quality of paper to be produced and the capacity of the paper machine (which may be different from that of the pulping plant).

IV.2 Market size

As indicated earlier, demand for paper products in a large number of developing countries does not generally justify the establishment of large pulp and paper mills unless the excess production can be exported. Furthermore, some paper grades consumed in these countries (e.g. newsprint) cannot be profitably produced in small mills. Therefore, profitable, large production of all grades of paper will require a fairly large domestic market or the possibility of exporting the excess supply to neighbouring countries i.e. the absence of trade barriers.

It may also be noted that even large countries may choose to establish both small and large pulp and paper mills, depending on the type of raw materials available and that of papers to be produced, plant location and other circumstances. India, for example, has promoted the establishment of both small and large plants, which cater to different markets.

IV.3 Quality considerations

The quality of paper products should be considered from two points of view: the characteristics of the paper produced (see Chapter II) and the uniformity of the output. Good quality paper - in terms of paper characteristics - may be produced in small mills with a minimum of mechanisation. Indeed, some of the highest quality papers can only be manufactured by small-scale mills (e.g. art papers, banknote paper). It may also be noted that the desirable stretch properties of sack-kraft paper may be difficult to achieve with large width paper machines.

Paper characteristics are also a function of the raw materials used for the production of pulp. Since most small pulp mills use agricultural residues, waste paper and other non-wood materials, the quality of the pulp may not be as high as that of the wood pulp produced by large mills.

The uniformity of the output is a function of the level of instrumentation and process control. Usually, large paper mills achieve a high level of product uniformity which is required by the markets of industrialised countries. A similar level of uniformity cannot be achieved by small paper mills which are not equipped with the necessary computerised control systems or other types of control instruments.

A paper producer should assess market requirements in terms of paper characteristics and product uniformity. In some cases, a low paper quality level may be acceptable in a segment of the market if the retail price is much lower than that of higher quality paper which is either imported or produced in a large mill. If an adequate level of product uniformity is required, efficient control systems may be added, at moderate cost, to the equipment of a small paper mill. Such controls have been added to relatively small machines (2 to 3 metres wide, and a capacity of 10,000 to 20,000 tonnes per annum) and have greatly improved product uniformity even in cases where the pulp is composed of waste-based materials.

The quality of waste-based papers can be enhanced through an improvement of the technology. It took years of research and development efforts to reach the current high quality level of wood-based papers. Similar research and development activities are currently being carried out to improve the quality of waste-based papers. For example, the quality of bagasse-based papers is close to that of wood-based papers whenever it is produced in small mills equipped with adequate control equipment. Similarly, research and development work has resulted in a significant improvement of the quality of straw-based printing and writing papers. Additional research is needed if high quality standards are to be achieved. Unfortunately, most research and development activities are carried out in developing countries where the bulk of waste-based paper is produced and these countries cannot afford to allocate sufficient funds for these activities.

IV.4 Capital investments

The capital investments considered in this section concern the paper mill only, and do not cover infrastructural costs (e.g. transport, electricity and water distribution) which are usually borne by the government. Those costs are considered in a following section.

Most developing countries are short of capital for agricultural and industrial investments. It is therefore very important to keep capital expenditures on development projects to a minimum, for example through the choice of appropriate scales of production and technologies.

Table I.1 shows the effect of alternative scales of production and technologies on capital investments in the paper industry. Estimates of these investments are provided for four types of paper mills: mini mills producing 500 tonnes per annum (TPA), small mills producing between 25 and 30 TPD and a large mill producing 300 TPD. It can be seen that capital investments for a total yearly production of 90,000 tonnes of paper are 15 to over 1,300 times higher for large paper mills (300 TPD) than for smaller ones. A preference for large rather than small paper mills could thus have important repercussions on investment in other sectors of the economy: some other investment must be cancelled or postponed if a large option is adopted.

Table I.1

Investment and foreign exchange costs

(Assumed production level: 300 TPD or 90,000 TPA)

(In million US$)

Kind of mill (daily output)	No. of mills required	Total investment cost	Investment cost per mill	Initial foreign exchange cost	Annual foreign exchange cost
Mini mill (1.7 TPD or 500 TPA)	180	19.8	0.11	10.8	4.14
Small mill (25 TPD)	12	57.6	4.8	10.6	11.1
Small mill (30 TPD)	10	100	10	47	13.95
Large mill (300 TPD)	1	150	150	90	13.56

Shortage of capital is not the only factor which militates against investment in a large paper mill. Two other factors are equally important. First, a large investment in a single paper mill will involve a high capital risk especially if it is the first such mill being built in the country. Investment in a number of small paper mills will considerably decrease the capital risk since a learning process takes place as the mills are being built over an extended period of time. Secondly, investment in large scale paper mills costing 100 to 150 million dollars will require direct government involvement since private local investors are unlikely to obtain sufficiently large loans from local banks. On the other hand most local investors should be able to secure the necessary funds for the establishment of small paper mills.

IV.5 Foreign exchange costs

The establishment of a paper mill will generally require two types of foreign exchange expenditures: initial expenditure for imports of the equipment and annual expenditure for imports of materials and spare parts. In some cases, foreign exchange is also required for the salaries of the foreign staff.

Most developing countries must postpone important projects because they cannot generate sufficient foreign exchange for imports of essential equipment and materials. Thus, scales of production and technologies which reduce foreign exchange expenditures to a minimum should be favoured. Table I.1 provides estimates of initial and annual foreign exchange expenditures corresponding to a total annual production of 90,000 tonnes of paper. It can be seen that a single 300 TPD paper mill requires two to nine times more foreign exchange for the initial procurement of equipment than do smaller mills. Annual foreign exchange expenditures are fairly similar for the large mill and for the two small mills (25 and 30 TPD), while it is considerably lower for the 500 TPA micro-mills.

IV.6 Employment effect

The generation of productive employment is one of the most important development objectives of many developing countries. Table I.2 shows that the establishment of micro-mills or small paper mills instead of a large plant should contribute significantly to that objective. As indicated in the table, small mills generate three to seven times more jobs than a large paper plant for the same output of 90,000 tonnes of paper a year. Furthermore, investment per worker is considerably lower for small mills. Thus, investment in these mills should help save scarce capital for other development projects which will generate additional employment. Furthermore, small paper mills, unlike large plants, do not rely very much on foreign expertise, if at all. Thus, most of the wages are spent locally and benefit the local economy.

Table I.2

Scales of production and employment generation

(Assumed production of paper: 300 TPD or 90,000 TPA)

Kind of mill (daily output)	Number of mills required to produce 300 TPD	Total No. of workers	Workers /tonne produced	Capital cost per mill (million US$)	Total capital cost (million US$)	Investment per worker (US$)
Mini mill (1.7 TPD or 500 TPA)	180	10 980	36.6	0.11	19.8	1 803
Small integrated mill (25 TPD)	12	5 640	18.8	4.8	57.6	10 212
Small integrated mill (30 TPD)	10	5 000	16.66	10	100	20 000
Large integrated mill (300 TPD)	1	1 500	5.0	150	150	100 000

IV.7 Infrastructural costs

Large pulp and paper mills, especially those which produce wood pulp, require an important infrastructure in such respects as means of transport and electricity and water supply. Most developing countries contemplating the establishment of a large paper mill must invest large sums of money for the improvement and expansion of their infrastructure whereas small paper mills may use the existing infrastructure. Thus, a comparison between small and large paper mills should take into account the differences in infrastructural costs although they are not explicitly included in the cost of the mill.

IV.8 Transport costs

The establishment of small paper mills should reduce transport costs because the raw materials (e.g. waste paper, agricultural residues) are collected nearby and the output is marketed locally. This is not so for large paper mills. The raw material - mostly wood - must be transported over long distances. The output generally exceeds local demand, so that part of the paper produced is usually marketed nationally, and must therefore be transported over long distances. The high transport costs are ultimately reflected in the retail price of the paper.

IV.9 Energy consumption

Since developing countries must import a large fraction of their energy, preference should be given to energy-saving technologies. Large paper mills, with the exception of groundwood-based mills are usually self-sufficient in power generation. This is not the case with small paper mills which must generally rely on power from an external source. A major cause of the under-utilisation and inefficiency of these mills is the unreliability of power supplies. Some new technological developments may partially solve this problem for mills producing 50 TPD or more. For the time being, however, large plants are more economical in terms of power consumption, although this advantage is partly offset by the additional energy needed for the transport of raw materials and output.

IV.10 Production costs

An important criterion for the choice of scale of production and technology in paper manufacturing is the retail price of paper which is, in turn, a function of unit production costs. The latter includes a fixed and a variable element. It was shown earlier that investment costs per tonne of paper produced are much higher for large plants than for small mills. Thus, unit fixed costs should be higher for large plants. Unit variable costs include the cost of labour and materials and that of energy. They are difficult to estimate without a precise knowledge of local conditions (e.g. in terms of wages, prices of agricultural residues). Thus, a general statement regarding the retail prices of paper produced by small and large mills cannot be made in this memorandum, especially since differences in quality further complicate the comparison. A few remarks on this matter can however be made.

First, in India small paper mills seem to be as competitive as large mills, and are currently supplying a growing share of the market. Secondly, retail prices are affected by the market structure. Thus, they should be much higher if paper is produced by a single, large plant in a monopolistic position than if it is produced by several competing small mills.

IV.11 Factors of interest to investors

There are several factors of particular interest to investors who may have to choose between different scales of production and production techniques. These are the gestation period, profitability and efficiency. These factors greatly favour small paper mills.

First, the gestation period of small mills is usually two years while that of large plants is at least five years. Secondly, experience shows that large plants seldom become profitable before the end of the third year of operation. Profitability is not always certain and is, at best, low. Small mills become profitable within a year and bring high returns on invested capital. Thirdly, the efficiency of large mills is generally higher than that of small mills. However, a well-organised structure is required to keep a high level of efficiency. In the absence of such a structure, small mills (which are easier to run and manage) can become more efficient than large mills.

IV.12 Rural and regional development

The establishment of several small paper mills should benefit rural areas and backward regions, whereas that of one or two large plants in urban areas will further aggravate income and employment disparities between urban and rural areas.

Small mills should create substantial backward and forward linkages including collection and transport of agricultural waste, establishment of small repair shops, manufacture of some of the equipment and marketing of the output. All these activities should improve rural incomes and slow down rural-to-urban migration. On the other hand, a few large plants, established of necessity in urban areas will generate few backward and forward linkages, will depend to a large extent on foreign expertise and will thus fail to promote technological self-reliance.

IV.13 <u>Environmental effects</u>

Environmental and pollution hazards are more damaging in the case of large plants than in the case of small mills. While the quantity of pollutants per tonne of paper produced is equal for both types of mill, more harm is done by large plants because large quantities of pollutants are disposed of within a limited area. On the other hand, the quantity of pollutants generated by isolated small paper mills is fairly low and can easily be disposed of without harm to the environment. Furthermore, treated effluents (containing lime) can be used as irrigation water, and thus benefit local farmers. Another advantage of small paper mills is that they do not generate as much air pollution as do large mills.

V. <u>GUIDELINES FOR GOVERNMENT ACTION</u>

The previous section provided information on the potential effects of alternative paper-making technologies, such as employment generation, foreign exchange savings or rural industrialisation. The importance of these effects will vary from country to country, depending, for example, on the prevailing socio-economic conditions and the local infrastructure. The weights assigned by public planners to these various effects in the process of evaluating alternative paper-making technologies and scales of production will, therefore, also differ from one country to another.

The formulation of an overall plan for the development of a country's paper industry requires the consideration of the following:

- the current and future local demand for various types of paper products;
- potential exports to neighbouring countries;
- the availability of various types of raw materials (see Chapter III);
- the current production, if any, of various types of paper products;
- the country's development objectives; and
- the socio-economic effects of alternative paper-making technologies and scales of production.

Once plans have been formulated for the development of the local paper industry in the short, medium and long term, there may be a need for government action in order to ensure the successful implementation of these plans. In particular, the government may need to formulate various policies and measures such as those applied by the Government of India in order to ensure the establishment of small paper mills. Depending on local conditions and circumstances, policies and measures may include the following:

- limited cash subsidies for small mills established in rural or backward areas; for example, the Government of India offers a cash subsidy which varies between 5 per cent and 7.5 per cent of the cost of the mill, subject to an upper limit of US$180,000;

- a ban on imports of specific types of paper products which are already produced locally;

- duty-free import of second-hand equipment for small mills;

- excise duty rebates which are a function of the size of the mill and of the type of raw materials used;

- exemption from excise duty on power consumption by small mills for a given number of years;

- favourable credit conditions for small mills;

- reduction of the proportion of the income assessable for tax, for a number of years after the start of the mill; and

- quotas on imports of some types of equipment.

The implementation of such measures requires a clear definition of the types of mills which may benefit from them. In some countries, such as India, plant capacity (in tonnes per day) is used as a criterion for the provision of various subsidies and other incentives to paper mills. For example, India has set up an upper limit of 30 tonnes of paper per day for the granting of various benefits to small paper mills. Some experts in this field are of the opinion that the capacity criterion may not be very efficient and could be advantageously replaced by other criteria such as machine size, speed and width. However, the choice of criteria need not be uniform: each country will need to identify those that are most appropriate in the light of local conditions and circumstances. It is, however, important to avoid the choice of criteria which may stifle the growth of the paper industry or increase production costs without compensating benefits for the national economy.

CHAPTER II

CHARACTERISTICS OF PAPER PRODUCTS

I.   PRODUCT RANGE

The range of paper and board products can be classified into types and grades: types relate to the paper or board machine design, while grades are a function of the end-use of the product and the choice of raw materials.

I.1.   Types of paper and board

Paper

Paper is generally defined as a single-ply, flat material, varying in density and material content according to end-use. It is traditionally produced on a Fourdrinier machine with an endless, moving wire-mesh screen. This principle of formation is still the most commonly used, and is particularly suitable for small paper machines.

Board

Board is defined as a multi-ply material which normally has a greater density than paper. It is most commonly produced on a vat machine where a series of rotating cylinders, of open design and covered with a wire-mesh, successively form layers of paper. These are added one upon another in a wet state until the desired thickness is reached, and are moulded together by pressing. The simple vat, which is speed-limited by centrifugal force, has been progressively improved in recent years in order to overcome the speed limitations. These improvements resulted in the development of the "former", or "dry" vat which is capable of maintaining high production levels and of producing various types of boards.

The Fourdrinier machine and the vat may also be used for the production of board and paper respectively. Board may be produced by the Fourdrinier machine by simply piling layers of material or by using a succession of wires for the forming of multi-ply board. Vats can also be adjusted to produce paper or single-ply board.

I.2. Grades

The types of paper or board can vary according to end-use. There are four distinct grades of paper and board.

Unfinished paper and board

Paper or board in this grade is not submitted to any surface smoothing process other than that naturally imparted by the press and dryer.

Machine finished (M.F.) paper and board

This grade of paper or board is obtained by passing the material through a calender, or vertical stack of chilled cast-iron rolls, before the paper or board is reeled. The paper acquires a smooth finish and a more regular thickness. Some board and paper machines have two calender stacks which yield a greater degree of smoothness.

Machine glazed (M.G.) paper and board

This grade of paper or board is obtained by pressing the moist material firmly against the surface of a drying cylinder. The moist paper adheres to the cylinder surface until it is dry enough for separation. A surface smoothness equivalent to that of the cylinder is imparted to the side of the paper or board in contact with it. Thus M.G. papers have a smooth finish on one side only. The drying capacity of the paper machine is limited to that of the M.G. cylinder which is larger than the usual drying cylinders.

Creped papers

The M.G. machine can be used to produce creped paper. This can be done with the addition of a scraping blade, or "doctor", which removes the paper from the cylinder before it is dry enough for natural separation. "Wet" creping can also be obtained by scraping from a press roll before normal drying. This technique is however less effective and more rarely employed.

Secondary finishes

The foregoing finishes are obtained as part of the machine process. It is possible to improve on the characteristics of paper or board through the use of secondary equipment. These secondary finishes do not, however, concern small mills, especially in the early stages of their establishment.

I.3. Types of pulp

Paper or board is produced from raw fibre sources which must be subjected to a pulping process for the production of stock. The latter is then processed in the paper or board machine and converted into the desired product. There are two major pulping processes:

Mechanical pulping

The fibre source is treated mechanically by grinding or "refining" through attrition between metal surfaces. The yield of this process is relatively high because only the solubles and a small fraction of "fines" are lost. However, the resultant paper is weak because the fibres do not bond chemically. The process is most suitable for wood-based paper or board, but has also recently been applied successfully to bagasse.

Chemical pulping

In this process the raw material is cooked, or digested, in the presence of chemicals. Thus, the individual fibres are separated from the natural, non-cellulose material which is taken away as a liquid suspension. The yield is lower than that for mechanical pulping, and is dependent on the degree of cooking and the nature and quantity of chemicals used. The latter vary according to availability, recovery potential and pulp characteristics required. Chemical pulp can develop strength and other characteristics by secondary mechanical treatment, and can be bleached to a high degree of whiteness, or brightness, by further chemical treatment.

Other pulping processes

In "semi-chemical" processes, a lower degree of cooking with less chemical is used to produce a higher yield and a pulp suitable for certain paper grades. In "chemi-mechanical" processes, higher yields are achieved by combining chemical and mechanical treatments. Mechanical or other high-yield pulps may also be bleached to a lower standard of brightness than fully chemical pulps because bleaching is merely an extension of the digestion process which reduces the non-cellulose content further than can practically be done by cooking alone. The consumption of bleaching chemicals in chemi-mechanical pulping is not economical if one wishes to reach the same level of purity as in chemical pulping.

II. PAPER AND BOARD GRADES

It would be beyond the scope of this memorandum to describe all the grades of paper which are manufactured. Description is limited to the most common grades of paper and board which may be produced in small plants operating in developing countries. It should, however, be made clear that any grade can be produced in these plants to any accepted standard of quality. Some grades, however, are more suitable than others.

II.1. Newsprint

Newsprint is made mostly from mechanical pulp although it may contain up to 30 per cent chemical pulp with a view to increasing its strength. It is relatively weak (40-55 g/m^2) and usually made of non-bleached pulp. It is essentially a cheap grade of paper in view of the fact that it is not intended to last long. It is normally produced in large plants with a capacity of 100,000 tonnes per annum (TPA) or more. The speed of production is relatively high and may reach up to 900 metres/minute in Fourdrinier type machines. In recent years, twin-wire machines have been developed to increase speed and efficiency while reducing the consumption of expensive chemical pulp.

Although it represents a substantial proportion of the total imports of paper and board by most developing countries, newsprint is not usually suitable for production by small paper machines. The power requirements are substantial (2,000 kWh/tonne or more), and are seldom available. The selling price is low (approximately US$450/tonne f.o.b.) and the added value small. However, small newsprint production may be competitive in the following cases:

- de-inking and recycling suitable local waste paper, possibly with the addition of imported waste paper;
- using the thermo-mechanical process if suitable energy and waste wood resources are available; and
- using the recently developed bagasse thermo-mechanical, or chemi-mechanical processes. This is, perhaps, the most promising field for developing countries with bagasse resources.

Generally, however, the small- scale production of newsprint is not appropriate to conditions prevailing in the majority of developing countries. Better returns can be obtained from the limited resources of fibres and capital available if they are used for the production of other types of paper.

II.2. Writing and printing paper grades

There is a large range of standards and prices for writing and printing papers. The various grades can, however, be integrated into four main classifications. All of these are eminently suitable for the small mill because the added value is substantial and there is usually a local market. The classifications are set out below, in ascending order of sales value.

Scholastic or educational papers

Papers in this category are produced for exercise books, notebooks, sketch pads and other articles used in educational institutions. The paper is bleached, but a high standard of brightness is not essential. It is relatively lightweight, (around 50 g/m^2) and need not be strong. Few additives are used. It is normally made on a Fourdrinier machine and machine-finished. Since the sales price must be as low as possible, almost any cheap fibre source (chemically processed) can be used provided it is bleachable to at least 60° G.E.[1] brightness (i.e. the newsprint standard). Waste paper of suitable quality can be added to the furnish (i.e. the blend of fibres and additives which make up the final paper sheet).

Banks

These papers are of better quality but light-weight, with a substance not exceeding 60 g/m^2. Brightness for white grades should not be less than 70° G.E., but papers of this grade may also be tinted. A three-stage bleaching is normally applied. The paper has many uses, such as for typing paper (principally for copies), printed books and leaflets. It is made on a Fourdrinier machine and machine-finished. It contains additives such as clay, starch or resin, to improve typing or printing characteristics.

Bonds

Bonds are similar to banks, but are of higher substance (above 60 g/m^2) and more durable. They are used as typing paper (originals), for printing or as stationery. They are produced on a Fourdrinier machine and machine-finished. They are normally white with a high brightness, although this category of paper also includes tinted grades. In order to satisfy

---

[1] This measure of brightness is named after the G.E. reflectometer, the instrument used to measure brightness.

the high quality standards required by the market for this type of paper, various additives are used to improve appearance, and the finish is size-pressed (an on-machine process) in order to improve writing or printing characteristics.

Art papers

Drawing paper, cartridge paper and water-colour paper are examples of this grade. The small mill competes advantageously in this field, particularly in developing countries where cotton is grown and can thus be added to the furnish for the production of high quality paper, including exclusive notepaper. A fraction of cotton, as low as 10 per cent, may significantly improve quality. Hand-made papers, made up exclusively of cotton, have a good export potential. Cotton waste is often added to the furnish although clean cuttings are more desirable. Cotton rags and linters, and fluff from the cotton mills are also employed where available. Apart from hand-made paper, the Fourdrinier machine is used to manufacture these grades of paper. In order to obtain high quality cartridge paper, the machine uses two wires which ensure similar appearance and characteristics on both sides of the sheet. Substance is usually high (80 g/m^2 and above).

III. WRAPPING OR PACKAGING GRADES

Wrapping paper standards are low in developing countries, especially since imports cannot be afforded while the need is great. Clean newsprint rejects can be sold for wrapping purposes at a higher cost than the original price. Most waste paper is also used as wrapping material before it finally ends in a paper mill for recycling. In developing countries, quality takes second place to availability. This offers a good opportunity for the small mill to engage in the production of this type of paper. The products made under this category have a surprising range of end-uses.

In the long term, better quality may be required. Small mills will probably need to improve their production technique in order to respond to demand for higher quality wrapping paper.

Sack kraft is not suggested for production by small mills because of the low added value and the long-fibre requirements. Substitute sack kraft is made commercially from bagasse, with 50 per cent imported pulp. Waste-based sacks are also made, but this process still needs further development. Because of the emphasis on strength and the limitations of long-fibre resources, prospects for the small mill are better in other grades.

Waste-based paper, combined with straw or bagasse pulp, is also used on M.G. machines for the production of "manilla" envelope papers for industrial use. The latter may constitute a profitable grade. Small mills producing writing or printing papers generally do not produce wrapping papers also. However, the same machine can produce a number of varieties of products and versatility is advantageous for a relatively small local market.

III.1. Boards (for packaging)

Boards are mainly used for packaging, and to some extent for such purposes as book covers, files and postcards. There are two major grades used in packaging: linerboard and boxboard.

Linerboard

This material is principally used for the interior or exterior layers of corrugated cases. In large plants, it is made in two plies from pure, unbleached softwood kraft, which is not suitable for production by small machines. If suitable wood is available (which is not often the case in developing countries), better use could be made of it. A lower grade substitute to linerboard, popularly known as "chipboard", can be produced from waste paper on small machines. A better grade, known as "test-jute liner", having only a top ply of kraft pulp, is also made in quantity by manufacturers in developed countries which have large supplies of waste paper but no suitable wood. Linerboard can be bleached: it is then known as "food-board" because it is commonly used for food packaging. The small mill is not suited for the production of this grade since the added value is too small.

Folding boxboard

This material is normally used for various qualities of cartons or boxes. It is of multi-ply nature (up to 9 plies), and is normally made on vat machines. The "former" is also increasingly used with a view to improving quality or (in large mills) speed. Normally, boxboard has a white or coloured top layer of chemical pulp which accepts printing, the remaining layers being made of waste paper. Superior grades may use high quality pulp for the bottom layer while the best quality grades may incorporate more than one layer of chemical or mechanical pulp for the top or bottom surfaces. Boxboard is often M.G. finished in order to provide a superior finish to the top side. Alternatively, it can be machine finished and coated (with a simple on-machine unit) with a clay-based coating mixture. For greater capacity units a secondary coating machine is used. Boxboard is suitable for production in developing countries, the scale being mainly dependent on the quantity of waste which can be obtained. In view of the intense competition for waste paper and the relatively small, overall supply of this material, the scale of production in developing areas may need to remain small for some time to come. Boxboard is a profitable product, and relatively simple to manufacture.

III.2. Wrapping paper

The great limitation for wrapping or packaging paper in developing countries is the availability of suitable fibre sources since strength is desirable for most grades and long-fibre material is seldom available in quantity. However, the small mill can make wrapping or packaging grades under favourable circumstances and can alleviate, to some extent, the need for imports.

III.3. Corrugating medium or fluting

This material is used for the corrugated, inner layer of paper between an outer and inner layer of linerboard to produce corrugated cases. In this instance, stiffness rather than tensile strength, is the desired characteristic. Traditionally, the material was made from straw. At present, use is made of semi-chemical hardwood which is being gradually replaced by waste paper even in developed countries. This material can be made competitively in small mills in countries where enough waste paper is available. Most of the small mills in India concentrate on this product. In rural areas straw is still being used. The added value is low, but so are capital and operating costs. Fluting is usually produced on a simple Fourdrinier machine and is unfinished. The quality of waste-based fluting can be improved by adding starch through a size press· However, this is seldom done in developing countries because capacity is reduced. In developed countries, fluting is produced in two or three plies on "former" type machines. It is claimed that this technique improves quality.

Bagasse or straw, cooked semi-chemically, makes good quality fluting when mixed with approximately 25 per cent of waste paper. This is a promising field for small mills where a local market exists. The material is normally dyed to resemble kraft paper, and is used for wrappings as well as for fluting in developing countries. The substance varies from 90 g/m^2 to 160 g/m^2. The heavier grades may also be used as file covers or board substitutes.

III.4. <u>Bag paper</u>

This paper should ideally be made from softwood pulp in order to improve its strength. However, other acceptable grades made from non-wood indigenous sources can be produced in small mills. Old jute, or gunny sacks, are particularly good for the production of paper of this type since they have long fibres and are relatively inexpensive. The M.G. machine is particularly suitable for the production of bag paper. This may explain the strong demand for small, second-hand machines of this type in India. The paper is normally lightweight ($30g/m^2$ to $45g/m^2$), but heavier grades are also produced. Bag paper and decorative wrapping papers are also produced in bleached or tinted grades. The long-fibre material used in the production of these papers may include refuse cotton, kenaf, hemp or sisal.

III.5. <u>Tissue grades</u>

Tissue papers are mainly produced in two grades: toilet and facial. The same machine may also produce paper towelling. Toilet tissues can be "hard" (i.e. uncreped) or "soft" (various creped varieties). They are most commonly made on M.G. machines, while the forming section can be a Fourdrinier machine or a cylindrical "former". Tissue products represent, perhaps, the best example of papers which can be profitably produced by small mills. The mills must be located near a population centre because the products are bulky, and transport costs can be excessive over long distances. The added value is high, but production is limited, per machine, to the drying capacity of one cylinder. In order to ensure softness, furnish is made of sulphite pulp. Tissue mills normally operate on imported pulp. Nowadays, a substantial proportion of selected waste is added to the furnish. Even in relatively developed countries, tissue mills with a capacity of 5 to 20 tonnes per day can operate profitably. For developing countries, however, tissue products are luxury goods, the market being restricted to hotels or supermarkets catering mainly for tourists or high-income groups. Thus, demand may not be high enough to justify production even on a small scale.

Where a waste-based M.G. mill exists in or near an urban centre, it may be possible to allocate part of the capacity to cosmetic or toilet tissue production. Bagasse pulp is a good material for the production of tissues. It can be used in proportions of up to 80 per cent of the furnish.

IV. TESTING OF PAPER AND BOARD

The essential characteristics of all paper and board products vary according to the end-use of the product. There are no defined international standards for any particular grade. In developed countries, standards may be set by government for example through postal authorities for packaging material. Even in such cases, however, the standards apply to the product directly and only indirectly to the basic materials from which the product may be made.

There are international standard methods for measuring and quantifying the qualitative characteristics of paper or board. In the interest of product uniformity (itself a quality), these should also apply to the small mill. The main characteristics and tests are briefly described below.

IV.1. Substance

Substance is defined as the weight per unit of area, usually in grams per square metre. It is measured by weighing a standard-size piece of the material on a specially calibrated balance. Writing and typing papers are normally sold, ex mill, by weight, but the customer expects a given number of sheets for a specified weight. If the mill does the cutting the product is sold by area, although it may be invoiced by weight because the customer will expect the correct number of sheets per ream or other specified package. It is therefore important for the mill to control substance because the customer is likely to complain if the weight is too low or the number of sheets lower than expected. Without adequate control, the only way to avoid complaints is to increase the weight without charging the customer. Even this method may not solve the problem if prepared boxes are used which cannot accommodate the correct number of sheets. Mills can lose substantial revenue in the absence of good substance control. This is why the acquisition of consistency regulators is generally justified and more than offsets the investment cost.

IV.2. Caliper

The determination of thickness, or caliper, is also required for board and some grades of paper because it indicates the strength of the material and can affect the performance of the carton or of the printing machine. It is measured by a specially calibrated dial micrometer. Caliper control is obtained by a combination of stock preparation, pressing and calendering.

IV.3 Testing of other paper and board characteristics

Burst

The Mullen test (TAPPI Method No. T403)[1] is used for the determination of bursting strength of paper by measuring the hydrostatic pressure (in lbs/sq. in.) required to rupture the material when pressure is applied at a constantly increasing rate to a circular area of 2.5 to 5 cm in diameter. It is a required test whenever strength is an important property of the paper. The test provides an average of both the longitudinal and transversal strengths.

Tear

TAPPI Method No. T414 is the accepted testing method for measuring the average force in grams required to tear a single sheet of paper after the tear has started. The standard Elmendorf tester is the most commonly used instrument. This test is applied mostly to wrapping paper.

Breaking length

TAPPI Method No. T220 is the testing method for measuring the length of a uniform strip of paper which will break under a certain tension when the strip is suspended by one end. A tensile strength and stretch tester is used for this test which quantifies the intrinsic material strength. The same instrument can be used to measure tensile strength or stretch.

Compression strength

This test is designed for papers where resistance to crushing is a desirable property (e.g. corrugating medium). Two main tests are used: Flat Crush, TAPPI Method T808, and Ring Crush, T472. In the former test, a defined strip of paper formed into a cylinder is subjected to a compressive force until buckling occurs. In the latter test, a circular corrugation is formed in the paper and then subjected to compression in order to determine the level at which it collapses.

Brightness testing

In this test a comparison is made between the amount of light reflected by the sample and that reflected by a standard paper under similar conditions. A reflectometer, of which there are several makes, is used for this test. The best known instrument is the G. E. reflectometer from which the G.E. scale originates.

---

[1] More details on the testing methods described in this section may be found in publications by the Technical Association of the Pulp and Paper Industry (TAPPI), 1, Dunwoody Park, Atlanta, Georgia 30341, United States.

Freeness testing

This test determines the freeness, or drainage characteristics of pulp. Two instruments may be used: the Canadian freeness tester and the Schopper-Riegler tester. Each has its own scale, but conversion from one to another can be made.

Smoothness

This test quantifies the surface finish of paper and measures air leakage under the rim of a disc placed on the surface of the paper. Air is introduced at standard pressure into the enclosed area. TAPPI Method No. T479 uses three tests: Gurley-Hill, Bekk and Williams.

Water absorbency

There are several recognised tests for this characteristic which is influenced by rosin sizing. The Cobb test is most commonly used: it measures the quantity of water absorbed after the paper has been subjected to water at constant pressure over a standard area for a specified time. Other tests float the paper on water, an indicator dye on the paper surface being used to measure the time taken for the water to pass through the paper.

Opacity

The importance of opacity is in preventing "show-through" (i.e. printing on one side of a piece of paper showing on the other side). The measurement of opacity is somewhat arbitrary. Several tests and instruments are used to measure this characteristic. The most commonly used instrument is a photometer with a divided field. It measures light reflected from the sample backed by a standard black background and compares it with light reflected from a similar sample backed by a standard white background. The comparison is expressed numerically, as a percentage.

The foregoing tests are the most common ones, and are likely to be required by a small mill. There are other tests for more specialised characteristics such as fold, pick, abrasion, ash content and grease penetration, but a small mill with limited resources can confine testing to the minimum that is still consistent with quality control and customer satisfaction. In addition to testing equipment it will need a drying oven, balances, a microscope, and a laboratory bench with fittings. It is important that tests be carried out as a matter of routine and the results analysed in order to improve quality and save materials. For example, it is important to check the moisture content of raw materials for a more meaningful costing of the output.

CHAPTER III

RAW MATERIALS AND ADDITIVES

I. FIBRE SOURCES

There are many fibres that can be used for paper manufacture. For small mills, there should be enough raw material that can be processed economically, taking into consideration the price of the material, collection and transport costs, expected yield and the cost of processing. By-products from other industries are the most desirable sources of material. Natural vegetation with no other outlet can also be used whenever wages are low enough to justify the harvesting and transport of the material. As already mentioned, there must be an assured local market for the end-product.

I.1. Short-fibres

Straw

Farmers have no use for between 5 and 10 per cent of all the straw they produce (e.g. from wheat, rice, barley), and it is commonly disposed of by burning. It is the most abundant source of fibres for developing countries. The characteristics of this material differ according to origin and climatic conditions. However, differences do not require adjustments to the production technique or restrictions regarding the choice of paper product. The average fibre length is 1.5 mm; the pulp is consequently weak and needs reinforcement. When the straw is chemically or chemi-mechanically cooked and bleached, yields of between 30 and 45 per cent can be obtained. Brightness may reach 72° G.E. Straw can be used as the major constituent of paper (e.g. 60 to 70 per cent for writing and printing papers). A higher yield (up to 70 per cent) can be obtained from chemi-mechanically cooked and unbleached straw. The pulp obtained can be used, with the addition of 25 to 30 per cent waste paper, for the production of corrugating medium. The addition of up to 50 per cent of long-fibre support is needed for the production of various grades of wrapping paper. The scale of production may vary between 15 and 250 tonnes per day. Chemical recovery, from fully bleached pulp mills, is economic from a level of about 25 tonnes per day capacity.

Bagasse

Where sugar cane is grown, bagasse is the most satisfactory and versatile source of fibre. The fibre is longer than that of straw, averaging 1.7 mm. Bagasse produces stronger pulp than straw and needs less long-fibre support (from 25 to as little as 10 per cent for paper of heavy substance). It bleaches well, attaining over 80° G.E. brightness with a standard three-stage bleaching. Yield from de-pithed[1] bagasse is approximately 45 per cent for bleached grade pulps and 70 to 75 per cent for unbleached, chemi-mechanical grades. However, 25 to 30 per cent of the material as received from the sugar mill is pith which should be removed before pulping. Thus, the overall yield from bagasse that has not already been de-pithed is lower. Bagasse produces excellent writing and printing papers, good corrugating medium (with 25 per cent waste paper support), and reasonable wrapping paper (with up to 50 per cent long-fibre support). Recently, high-yielding chemi-mechanical and thermo-mechanical processes have been developed for the production of newsprint. The refining requirements for these processes are low, averaging about 25 per cent of those for equivalent short-fibre wood pulp. Where fuels other than bagasse are available for the sugar mills, the scale of production of bagasse pulp can be as high as 150 tonnes per day per pulping unit for bleached grades, and higher for unbleached or newsprint grades. In the absence of substitute fuels, there should still be a surplus of bagasse to partly supply small paper mills.

Maize stalks

The properties of maize stalks are similar to those of straw. However, their collection and storage pose some problems especially in view of their high moisture content at picking time. These problems make maize stalks unsuitable as a basis for large production. In low wage countries, small-scale production can be carried out profitably.

---

[1] Bagasse from which the pith, a spongy cellular tissue, has been removed after the processing of sugar cane in sugar mills.

Bamboo

The fibre length of this material averages 2.7 mm. Consequently, the strength of the fibre is such that it can be used without addition of other fibres for the production of paper furnish. It is most commonly used for bleached grades of paper, with yields of 40 to 45 per cent. Preparation and bleaching are more difficult to achieve than for straw or bagasse, but the techniques are now well known. Bamboo is one of the more valuable sources of fibre in India where it is used for the production of all grades of paper, including newsprint. However, because there is not enough bamboo available, it is used on only a limited scale.

Other grasses, reeds

Naturally growing vegetation can be used whenever it is available in sufficient quantity and labour costs are low. Esparto grass and reeds for example, are good short-fibre sources. They can be used in mills producing up to 150 tonnes per day if the raw material is available in sufficient quantities. The production process is similar to that for straw, but preparation may require a special treatment.

I.2. Long-fibre materials

For most grades of paper, there should be access to a sufficient quantity of long-fibre materials in order to obtain the required paper strength.

Cotton

The cotton fibre is 25 mm to 32 mm long and contains almost pure cellulose: thus, it is a very good material for paper manufacture. However only cotton waste or process residues are used for paper manufacturing in view of the high value of cotton for textile production. Furthermore, the price of waste and residues is so high that they can be used only for the production of high grade papers, such as hand-made and art papers. A small proportion of cotton fibres is also used as a reinforcing constituent in the production of writing and printing papers.

Pulping and bleaching chemical requirements are low (single-stage bleaching is sufficient). Yields are relatively high, depending on the fibre source. However, beating energy requirements are also high.

Cotton is available to the papermaker in several forms of waste, which is ranked in terms of desirability and cost as follows:

- clean cotton cuttings from the textile mill;
- first cut linters;
- mill-run linters;

- second-cut linters; and
- fluff, from cotton mill exhaust systems.

The latter grade may not be suitable for bleaching, but is usually very cheap and can be used for wrapping papers. Cotton-based pulp mills are often small.

Rags

Rags from which woollen and synthetic materials have been excluded may be used for paper-making. In developing countries, they are likely to be of predominantly cotton stock. The sorting of rags includes grading by colour because the latter affects bleaching requirements. The preparation stages include chopping, dusting and washing.

Flax

Seed flax tow, a residue from the manufacture of linen, can be used for paper production. The fibres average 25 mm in length, and the pulp is used for the manufacture of specialty papers such as cigarette paper. The yield is around 30 per cent. Bleaching is difficult because the material contains both bast fibre and woody shivers with different cooking rates. The strength properties of flax tow make it a good support fibre for unbleached wrapping papers.

Hemp

Manilla hemp is obtained from Philippine banana stalks. Paper is seldom made directly from banana stalks because the fibre content is low and the removal of the non-fibrous pulpy element is costly. The manilla hemp pulp used consists almost exclusively of old ropes or manilla tow from ropemaking factories. The fibre length averages 6 mm. The yield is 50 per cent for bleached grades and up to 65 per cent for unbleached grades. Processing is similar to that of cotton waste. Bleaching is carried out in a single stage because true whiteness cannot be obtained, and the aim of bleaching is mostly to obtain a lighter natural colour. There are other hemps, such as sisal and Benares hemp. These fibres do not have all the characteristics of manilla hemp, but are nevertheless suitable for the production of wrapping grade support. Hemp is also pulped for specialty papers such as cigarette, carbon or condenser papers.

Jute

Jute is available to the papermaker in such forms as used burlap cuttings, old sacks and hessian. The preparation and digestion stages are similar to those for cotton waste, but lime and soda ash are sometimes used for the cooking of the material instead of caustic soda. Jute does not bleach well: it is used mainly as a support fibre for wrapping paper, or for some specialty grades in which toughness and durability are required characteristics. Yield is around 60 per cent and the fibre length averages 2 mm.

Other long fibres

Kenaf is a promising long-fibre source, with a fibre length averaging 2.6 mm. Preparation of this material is difficult because it is not available as an industrial residue. Thus, the whole plant must be processed including "retting" and the removal of non-fibrous material. The yield is low but the fibre is excellent. It also bleaches well. The fibre properties are attracting attention and development work is currently under way. If an economic process can be developed, kenaf could become an important source of long fibres for developing countries in addition to being used as animal feed.

Sabai grass is used in India for unbleached wrapping or packaging paper or board support fibres. The average fibre length is 2.1 mm and the yield around 35 per cent.

I.3. Waste paper

Waste paper can contribute up to 30 per cent of a developing country's paper needs. The best quality papers are obtained from the waste of printing houses. The bulk of waste paper comes from urban areas where consumption rates are the highest. In developed countries, waste-paper collection is organised by merchants through municipalities or charitable organisations. In developing countries, collection is seldom organised on a large scale but little is wasted. Small entrepreneurs collect waste paper from wherever it can be found and sell it directly to the paper mill. Sorting for the removal of non-paper material and classification of the waste is done at the mill. Classification may involve the separation of brown from white paper, white unprinted from white printed, heavily printed from lightly printed, and so on. This is economically feasible in view of the low wages of unskilled workers in developing countries.

Appendix II provides the European list of the standard qualities of waste paper.

I.4. Wood-waste

Sawmill wastes can be used as a fibre source in addition to other materials. However, their greater value as a fuel for the sawmill or other consumers limits the availability of this material. Another restriction to the use of this material is the need to apply a modified processing method.

I.5. Imported materials

If long-fibre pulp of suitable quality cannot be produced economically from local materials, a mill can still operate profitably by importing 20 to 30 per cent of its total fibre requirements. This applies, in particular, to straw-based or bagasse-based mills which may use up to 30 per cent of imported bleached kraft. Other imported materials may include unbleached sulphite pulp or various grades of waste paper. The further processing of such imported materials may, in many cases, be much more cost effective than the use of local materials such as cotton waste.

II. CHEMICALS

Chemicals are used in both pulp and paper mills and most developing countries often import the main ones, which are described in this section.

II.2. Chemicals for pulp mills

Caustic soda (NaOH)

Caustic soda is, with few exceptions, the cooking chemical used by small mills. It is available in solid form or in liquid form at around 50 per cent concentration. The amount used varies between 5 per cent of raw material inputs for cotton, hemp or jute, to 12 and 14 per cent respectively for straw and bagasse. Chemi-mechanical processes require less caustic soda (between 5 and 10 per cent) than bleached pulps. Caustic soda is also used in the bleaching process for the caustic extraction and hypochlorite stages.

Lime (CaO)

Lime is used in some small mills for low quality cooking of such materials as jute waste and inferior rags. It is also used as a constituent for calcium hypochlorite (see below). It is often produced at the mill through the burning of limestone in a simple open burner or kiln. It is also used as the causticising agent for chemical recovery.

Chlorine

Chlorine is used in the first stage of the three-stage bleaching process. It is also used in the last stage in the form of calcium or sodium hypochlorite. Overall consumption varies between 7 to 9 per cent of raw material inputs for three-stage bleaching. It is normally supplied in a compressed liquid form in metal bottles, and used as a gas.

Hypochlorite

Hypochlorite is used for single-stage bleaching or at the final stage in the three-stage process. It may be purchased in powder form, as calcium hypochlorite, or in liquid form as sodium hypochlorite. It is often manufactured at the mill by adding chlorine to slaked lime for the production of calcium hypochlorite, or by adding chlorine to caustic soda for the production of sodium hypochlorite. The latter is used whenever lime is not available locally, or when caustic soda is supplied in liquid form.

II.2. Chemicals for paper mills
Alum

Alum, or aluminium sulphate, is generally used for pH correction. It is added to the paper stock in order to increase its acidity to approximately pH 5. It is an essential additive for good refining, adequate sizing, or rosin dispersion. It also affects press performance by reducing the tendency of paper to adhere to the top press roll. It is most commonly supplied in solid block form, but should be dissolved in water and used in liquid form at a constant density. Some small mills add crushed alum to the pulp during the beating. This is not recommended, however, because alum is highly corrosive. This chemical is also used for water treatment in order to induce flocculation. Average consumption is 5 to 7 per cent by weight of paper produced.

Rosin

Rosin is commonly used for "sizing" paper. It improves printing or writing properties by controlling ink penetration. Raw rosin is available in solid form. It must be brought to a liquid emulsion before use. Small mills in developing countries normally produce the emulsion by cooking the resin with soda ash or a mild caustic solution. It is more common for large mills in developed countries to purchase ready-prepared rosin in drums.

Starch

Starch is used to improve the quality of writing and printing papers or the stiffness of waste-based corrugating medium. It is most effectively applied, in liquid form at about 30 per cent concentration, to the surface of the paper through a size press. However, the drying capacity of the machine is reduced unless additional cylinders are added. Therefore, this method of application will only apply to the small mill which produces paper according to the customers' specifications at an agreed price. Starch must be cooked and prepared in liquid form before use. Ready-prepared starch is available and used extensively in developed countries but is not likely to be used by the small mill in developing countries. Starch can also be used in a less effective manner, as an additive to the paper stock. From 2 to 10 per cent by weight of paper can be used, depending on the quality to be achieved. It may be noted that because starch is produced from various food sources (e.g. potato, rice, maize), its use in large quantities for industrial purposes may not be feasible in countries with limited food resources.

China clay

China clay, or kaolin, is used extensively in the production of writing and printing papers in order to improve opacity, brightness and finish. Up to 20 per cent of clay is used for some grades of paper. Where clay is cheaper than pulp, costs may be reduced by using as much clay as technically feasible. Suitable clay, however, is rarely available in developing countries and imported bagged clay can often be more expensive than the fibre. The addition of clay tends also to weaken the paper, and may consequently require more costly long-fibre support material. In general, the addition of clay should not exceed 8 to 10 per cent of total material inputs. It can be added in powder or slurry form to the pulp when it is being prepared (beaten).

Talc

Talc, where available, can be used as a substitute for clay in similar proportions. However, the pH of the paper stock must be close to the neutral point since an acid stock can dissolve the talc.

Other chemicals

There are a number of other additives which may be used in paper production, including dyes for tinting, titanium oxide to improve brightness, optical bleach, retention aids, muriatic acid for cleaning and scale removal, detergents and wet-strength resins. Small mills in developing countries may not always be able to use these materials since most of them must be imported, often at high cost.

CHAPTER IV

PULP PRODUCTION TECHNOLOGIES

A comprehensive description of all existing pulp production technologies is outside the scope of this memorandum. This chapter therefore provides a concise description of the technology most likely to be adopted by small mills in developing countries since these are the main focus of this memorandum. Thus, mechanical and thermo-mechanical pulp production techniques are not described for the following reasons. Mechanical pulping requires large capacity grinders which considerably exceed the production limit of small mills. Furthermore, suitable species of wood are seldom available in developing countries, and power requirements for grinding are fairly large. Similarly, the production of thermo-mechanical pulp is mostly appropriate for medium-size mills in view of the capacity of the special refiners used for the production of this type of pulp. This chapter will therefore focus on the production of chemical and chemi-mechanical pulps. A brief description of semi-chemical pulping will also be provided.

I. CHEMICAL PULP PRODUCTION

Batch cooking in spherical digesters is used to a large extent by mills with capacities of 50 tonnes per day or less. Above this level, continuous digesters are preferred for economical and technical reasons.[1] The chemical cooking agent for small mills is, with rare exceptions, caustic soda because it is more likely to be available, is simple to apply and can be recovered in mills with capacities of 20 tonnes per day and above. This section will therefore focus on chemical pulping based on that technology. Figure IV.1 shows the main processing stages of the chemical pulping of straw. They include the following:

---

[1] A continuous screw-type digester is usually less expensive and requires less space than a multiplicity of spherical digesters. Furthermore, it reduces cooking time and yields a more uniform pulp quality.

- 38 -

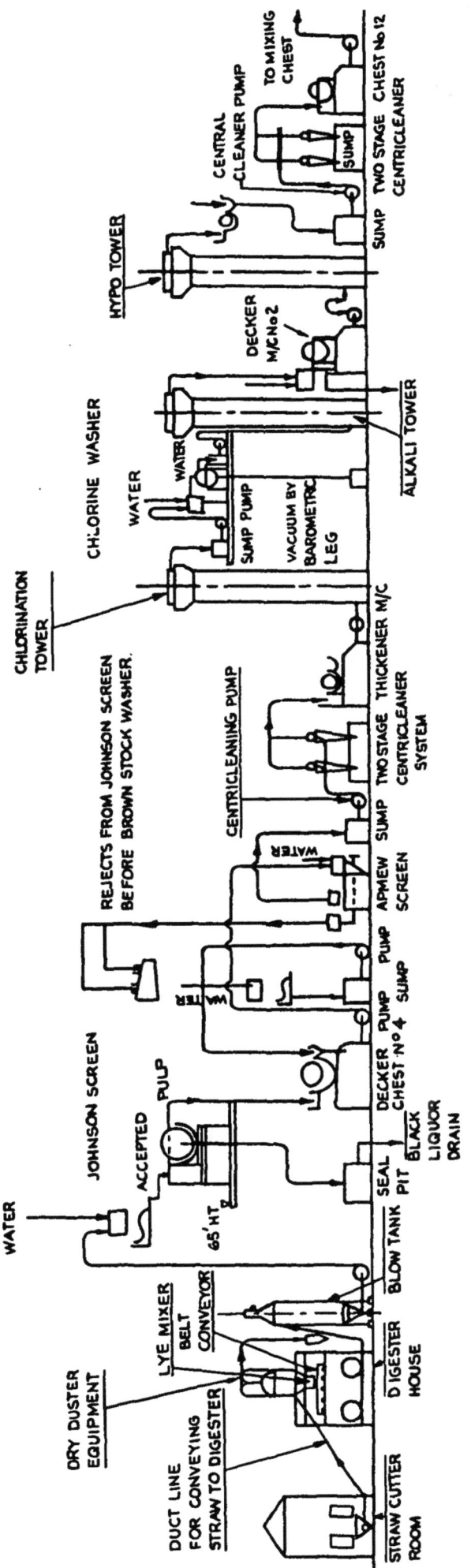

Figure IV.1

Flow sheet for straw pulp

- collection and storage of raw materials;
- fibre preparation;
- digestion (batch or continuous);
- washing;
- screening;
- bleaching;
- stock preparation, including the following: use of a hydrapulper, beater or deflaker; refining (use of conical or disk refiners); screening; and cleaning;
- chemical recovery; and
- effluent disposal.

These processing stages are described in the following sections of this chapter.

I.1. Collection and storage of fibres and waste paper

I.1.1   Collection and storage of fibres

The first processing stage consists in the collection, storage and preparation of the fibre source. Agricultural residues (e.g. straw or bagasse) or naturally growing vegetation (e.g. reeds or grasses) are generally available on a seasonal basis. Thus the year's supply of fibres must be collected over the harvesting period and stored. Since the production of one tonne of bleached pulp requires 4 to 7 tonnes of fibres (depending on the raw material adopted), a large storage capacity will be necessary even for a small mill. The baling of straw is fairly common while bagasse is baled only when it has to be transported over long distances.

Straw bales are formed into stacks which can be held in stock over the period required without significant deterioration if properly protected against rain. The density of baled straw is around 10 to 12 cubic metres per tonne. Stacks up to 14 metres in height and 2,000 cubic metres in volume are usual. They should be widely separated in order to avoid the spreading of fire and the gaps between them should be kept scrupulously clean for the same reason.

Bagasse can be stored in a similar manner but modern practice is to use bulk storage: moist bagasse is stored in bulk in order to minimise storage requirements. It is stored in piles 30 metres high in order to reduce the risk of fire. Exposed surfaces can catch fire and should therefore be kept wet by spraying during protracted dry spells. Good storage management is

very important. The aim should be, whenever possible, to maintain the same
storage time for all materials in order to ensure a good product unformity.
Mechanical handling with bulldozers, forklifts or front-end loaders cannot
generally be justified for small mills in developing countries. Storage and
handling losses should not exceed 5 per cent, but are very difficult to
quantify precisely because of moisture variations.

Storage requirements for a small mill can be quite considerable. For
example, a mill producing 10 tonnes of paper per day with a mixture of 70 per
cent of straw pulp and 30 per cent of imported pulp will require approximately
16 tonnes of air-dry straw per day. A small mill may need to stock enough
straw for three months'production or about 1,600 tonnes of straw with a volume
of 16,000 cubic metres. If allowance is made for the space that must be left
between the stacks of straw, this amount will require a covered shed with an
area of approximately 2,000 m^2 and a height of 16 m.

I.1.2 Collection and storage of waste paper

Waste paper collected for paper-making may include printers' wastes,
packaging materials from industries and stores, household waste and paper from
government offices. It should be sorted according to various grades and
cleaned. In particular, care should be taken to remove impurities which have
a detrimental effect on paper quality and paper machine operation. Such
impurities include staples, rags (which can give rise to breaks on the paper
machine because they will not defibrate when mixed with waste paper), asphalt
and wax. The removal of some of the impurities may be carried out during the
sorting of the waste paper, while other impurities are removed at the furnish
production stage. The sorting of waste paper may be performed manually, with
or without the use of a slow-moving conveyor belt. Acquisition of the latter
will be a function of the daily volume of paper to be sorted and of the wage
level of unskilled workers. If wages are low, the acquisition of the
equipment may not be justified.

In order to facilitate sorting, waste paper from various origins should
not be mixed. The paper is spread on a concrete floor and transferred into
separate containers or baskets according to grade. Impurities are left on the
ground and then discarded, while the paper is baled and stored. Each grade of
paper should be stored separately.

The storage area for waste paper should be approximately half of that
needed for straw for the production of the same quantity of furnish. No

precise estimates are available regarding the productivity of unskilled labour for the sorting of waste paper since it is a function of the type of wastes being processed. For a small mill processing 1 tonne of waste paper per hour, 50 or more unskilled workers may be needed.

I.2. Fibre preparation stage

I.2.1 Straw preparation

Straw should be cut to a uniform length and dusted before passing to the digester. The removal of dust is required in order to ensure good paper quality. Dust losses of around 5 per cent can be expected. A combined straw chopper and duster is most commonly used. The straw is cut into pieces 4 to 5 cm long in order to facilitate the blowing of the chopped, dusted straw to the digester and to improve cooking uniformity. The chopped and dusted straw is transferred to the digester on a belt conveyor.

I.2.2 Bagasse preparation

Bagasse, as received from the sugar mill, has already been reduced to an acceptable size by the crushers. However, it contains about 30 per cent of pith which should be removed before transferring it to the digester. Pith does not produce pulp of any value, and requires greater amounts of chemicals; furthermore, it adversely affects the quality and drainage characteristics of the final pulp product. The general practice is to "moist de-pith" bagasse before storage because this process reduces storage volume requirements. De-pithing can also reduce production cost if the sugar mill agrees to burn the pith in its bagasse-burning furnaces. De-pithing also reduces the degree of storage deterioration since the pith contains sucrose which promotes fermentation. For better results, "wet de-pithing" is carried out after storage in order to clean the fibre and to remove additional pith which has become detached from the fibres during storage. "Moist de-pithers" are hammer-mills rotating within a screen through which the pith is rejected. "Wet de-pithers" are either similar units which process a slurry of bagasse and water, or separate agitated slurrying vessels followed by rotating or vibrating screens.

Bagasse may not be the major fibre constituent: for example, it may be used in conjunction with straw because the sugar mill is prepared to sell only surplus bagasse, or because straw is cheaper but insufficient in overall quantity. In such cases, the cost and power consumption of the de-pither may not be justified. This case should be thoroughly investigated because the decrease in operating costs and pulp quality are important variables. It is possible to obtain some degree of de-pithing through the straw chopper or to prepare the material by a separate dry screening and agitated washing operation as an alternative to the expensive and power-intensive de-pither.

I.2.3 Preparation of other fibres

Vegetable residues other than straw or bagasse may require different forms of preparation. Kenaf, for example, is "retted". This is a soaking process required to facilitate the removal of pulpy, non-cellulose material. Banana stalks require crushing for the same reason. Prospective investors should investigate the whole subject of collection, storage and preparation before authorising a project and ensure that the most suitable method of fibre treatment is used. The success of the project is largely dependent on good practice in these respects.

I.3. Pulping stages

I.3.1. Digestion

The process of digestion consists in cooking the prepared material in the presence of caustic soda to isolate the cellulose fibre from the lignin and other constituents of the straw. Digestion may be carried out on a batch or continuous basis. For small mills (e.g. producing up to 30 tonnes per day), batch cooking is preferred. Its advantages include low capital costs, simplicity and versatility. It is for example possible to cook straw and rags simultaneously in the proportions required. The materials in the continuous digester are screw-propelled through the cooking zone because stringy vegetable materials do not flow naturally and mechanical propulsion is therefore required to ensure constant feed.

Batch digester

The rotating spherical batch digester is the most effective and versatile unit for integrated mills producing a maximum of 30 ADT[1] per day of bleached, cultural papers. It also requires low capital investments if ordered from India where batch digesters are produced in large quantities in view of the high local demand. Recent quotations indicate that the price of digesters manufactured in developed countries is two to three times greater than that of similar digesters produced in India.

The advantage of the batch digester lies in its ability to pulp indigenous long fibre constituents such as cotton and jute concurrently with the major furnish materials such as straw or bagasse. With such equipment, it is, for example, possible to produce economically, in the same mill, a pulp made up of a combination of straw, bagasse, cotton and wood, if available.

---

[1] Air-dry tonnes.

The spherical batch digester has a diameter of approximately 3.8 m and a capacity of 25 m^3. It is made of mild steel, which is not affected by caustic soda. It is rivetted or welded in order to withstand steam pressures up to 10 atmospheres. The digester is mounted on bearings so that it can revolve slowly during the cooking cycle.

A typical straw batch cooking programme for the production of bleachable grade of pulp is as follows:

	Time in hours
- filling with straw and charging with cooking liquor	1.5 - 2
- raising to steam pressure (8-9 atmospheres)	1
- cooking at pressure	2 - 2.5
- blowing and emptying	0.5 - 1
Total	5 - 6.5

The output per digester is approximately 4.5 BDT[1] of pulp per day on the basis of four batches per 24 hours. Thus, six digesters are needed for the production of 30 ADT of paper per day. For higher outputs, a continuous digester is more economical in view of the high capital cost - including equipment, buildings and land - of a series of batch digesters providing an equal total capacity.

Caustic soda consumption for straw pulping amounts to approximately 10 to 12 per cent by weight of bone-dry straw input. The straw to liquor ratio is kept between 3:1 and 4:1. The charging liquor is made up from the caustic solution, some returned spent cooking liquor (termed "black" liquor) and hot water. Water consumption amounts to approximately 2.5 to 3.5 tonnes per bone-dry tonne of unbleached pulp. Cooking is carried out at temperatures of 160°C to 165°C. The yield of the pulping process - in the form of unbleached pulp - is 33 to 36 per cent of the quantity of straw fed to the digester.

The above processing method varies with the origin of the straw. Thus, rice or maize straw require adjustments of the above process which applies to wheat straw. Similarly, the process may vary for the same species of straw, depending on the climatic conditions in the growing area.

---

[1] Bone-dry tonnes.

Bagasse requires more caustic soda than does straw (up to 14 per cent), but yields a larger fraction of unbleached pulp (50 per cent on de-pithed bagasse). The cooking period can also be shortened.

Continuous digesters

Continuous digesters are more economical for the large-scale production of pulp, and allow a better control of the pulping process. However, batch digesters may still be needed for the processing of indigenous long fibre materials (used as a support) unless imported softwood pulp is used instead.

There are two types of continuous digesters: the pressurised digesters (mainly Defibrator or Pandia units) and the atmospheric continuous digester (e.g. the SAICA type unit). The pressurised continuous digesters are more economic and effective for capacities exceeding 50 tonnes per day, but cannot cook such materials as cotton and old gunny sacks.

The SAICA atmospheric continuous digesters (see figure IV.2) are available in two sizes with a capacity of 40 tonnes per day and 80 tonnes per day respectively. They are especially designed for the processing of straw or bagasse, but can also process waste wood in view of their large volume and slow cooking period. They are currently used for the production of pulp for corrugating medium, but recent trials on bleaching grades are particularly promising. The need for continuous digesters for the production of 15 to 25 tonnes per day of bleachable pulp may be satisfied by the SAICA which is relatively simple to operate, requires a low capital investment, and can be partially manufactured locally.

Because they are compact, the cooking time of the pressurised continuous digesters is very short: 13 to 17 minutes. However, a short cooking time is not a virtue in itself; other conditions being equal, slow cooking produces pulp of better quality and requires slightly less chemical.

I.3.2. Blow tank

After cooking, the content of the digester is discharged, under pressure, into a blow tank where the pressure is released and heat recovered through a heat-exchanger. Water is added to reduce the pulp consistency to a level at which it can be pumped to the washing and cleaning section (see figure IV.3). The consistency of the pulp is between 2.5 and 3 per cent.

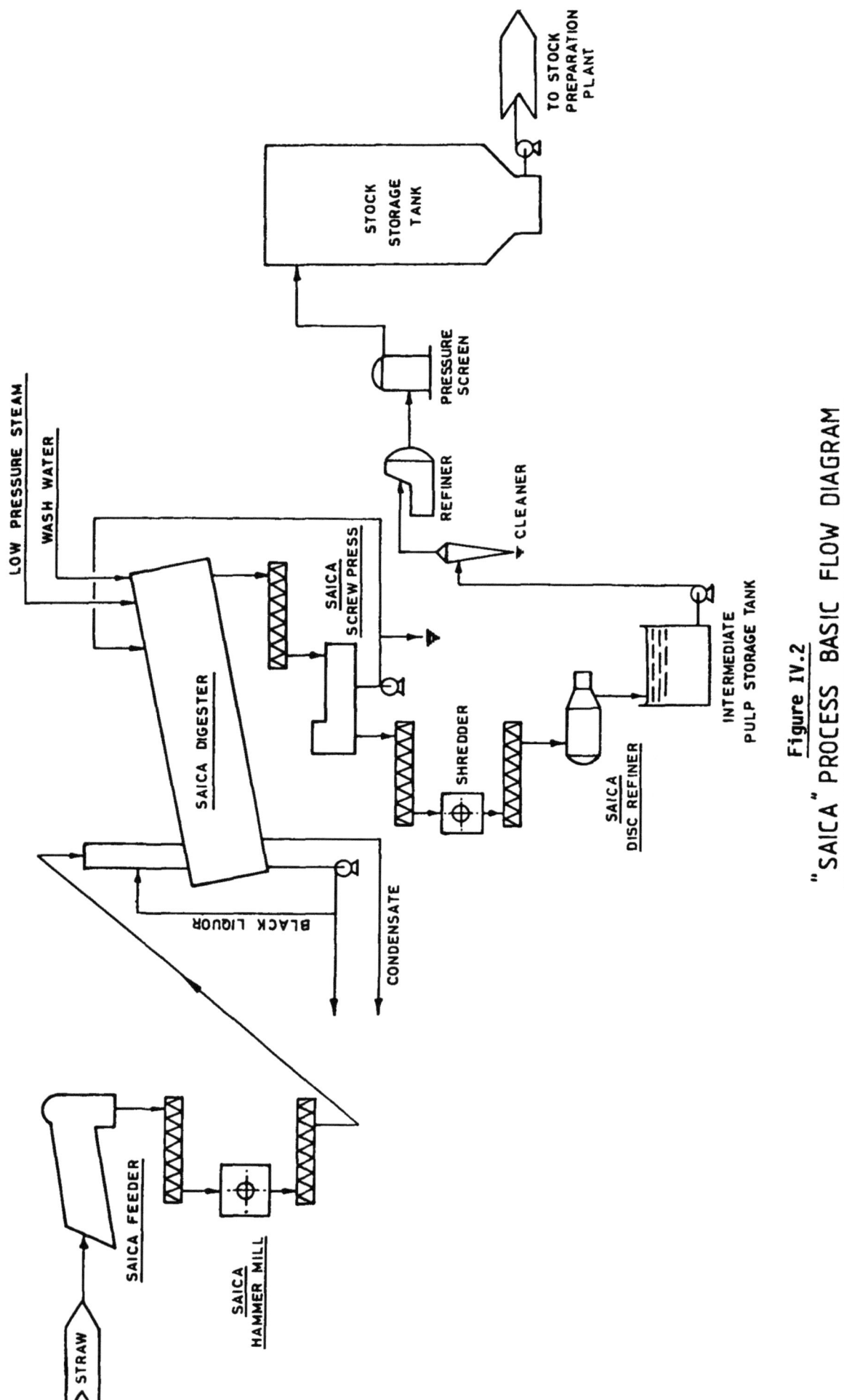

Figure IV.2
"SAICA" PROCESS BASIC FLOW DIAGRAM

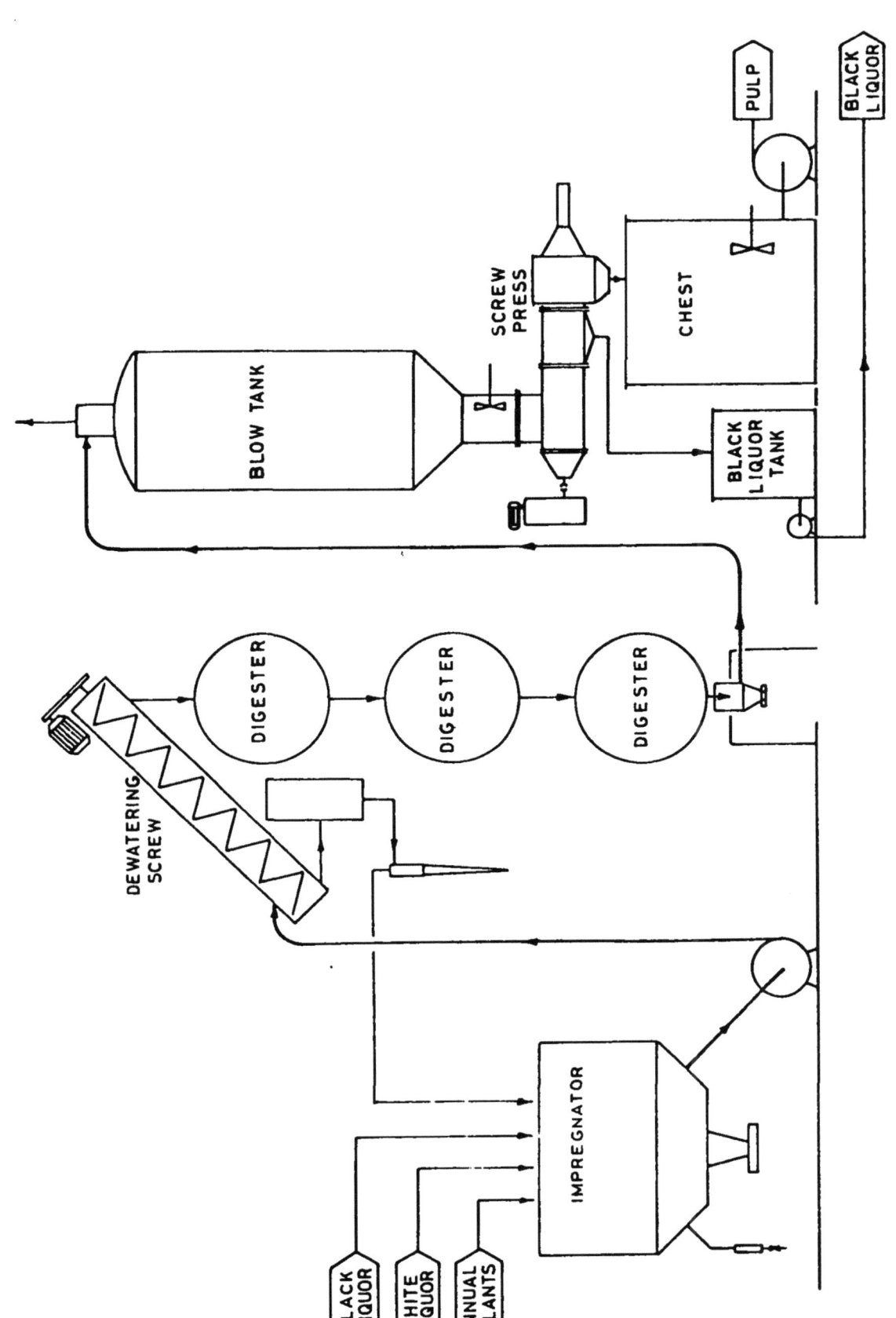

Figure IV.3
SULZER-ESCHER WYSS PULPING SYSTEM

I.3.3. <u>Washing</u>

From the blow tank, the pulp and accompanying black liquor is usually coarse-screened in order to remove uncooked particles (or shives), which should not exceed·5 per cent of the mixture. The rejects can be either returned to the digester for further cooking or mechanically processed to an acceptable condition by "refining". However, for mills of very small capacity the quantity of rejects is insufficient to justify investment in a refiner. In order to obtain good quality pulp, a three-stage, counterflow washing is used. This procedure should also reduce chemical losses if the mill is equipped with a chemical recovery unit. However, most mills of a capacity of 20 tonnes or less per day do not have such a unit, and use the single-stage washing process in order to reduce capital cost and power consumption. Improvements can be made later as capacity is increased or more funds become available.

Washers are open surface cylinders, covered with a wire mesh and rotating in vats. The unwashed pulp is fed to the vat and picked up by the cylinder. The liquor is drained away through the surface wire-mesh. In three-stage counterflow washing, fresh hot water is sprayed onto the pulp in the last washer for further cleaning. The drained-off water is then used to wash the pulp in the second washer and liquor drained from the latter is used to wash the pulp in the first washer. The black liquor drained from the first washer contains all the non-cellulose matter and chemical residues removed from the pulp. It is passed to a filtrate tank for disposal or chemical recovery except for a fraction which is used for the making of the cooking liquor supplied to the digester.

Washers can be either very simple and inexpensive or somewhat complicated and costly but more effective. The most satisfactory units operate under vacuum and require a special construction. Drainage is, in this case, improved and the on-going pulp contains less chemical or organic impurities. Washers used in small mills can be elevated in order to achieve the vacuum required barometrically. This procedure reduces capital costs and power consumption.

In principle, the composition of the black liquor removed is as follows. If the yield in the form of unbleached pulp is 40 per cent of the weight of straw fed to the digester, then the liquor will contain the 60 per cent of organic material removed from the straw. If the caustic soda added is 10 per cent by weight of incoming straw, the liquor will also contain that quantity of inorganic materials. In practice, the amount of both organic and inorganic constituents will be slightly less than expected because a small fraction will be retained in the washed pulp. Therefore, for each tonne of pulp produced, 1.75 tonnes of organic and inorganic material is produced and must be disposed of or processed in some way. In the liquor, the concentration of dissolved or suspended solids will depend on the quantity of washing water used. It varies between 5 per cent for ineffective washing to around 11 per cent for good washing. Thus, for each tonne of pulp produced, 18 to 40 tonnes of washing water can be required. The amount of solids in the liquor depends on the yield and the quantity of chemicals added, while the amount of water depends on the washer efficiency and, in some areas, on availability. However, if the washing is not good, undesirable matter goes forward with the pulp and adversely affects the bleaching quality and, ultimately, the quality of the paper produced.

I.3.4. Screening

The consistency of pulp leaving the washers varies from 5 to 12 per cent, depending on the effectiveness of the washer. The next stage is screening in order to remove particles unsuitable for the paper process. Screening is carried out at a consistency of 1 to 2 per cent and water must therefore be added to the incoming pulp.

There are several types of screens suitable for small mills. However, the best types are fully enclosed pressurised or centrifugal screens. After screening, the pulp should be thickened for "brown stock" storage. A decker, or thickener, is used for this purpose. It is similar to a washer but spray water is not added. The function of the decker is to remove surplus water. Storage at a decker density of 6 to 12 per cent is held in a high-density storage tower which, ideally, should have a capacity sufficient for holding 12 hours of production.

Some mills do not include this section in order to reduce capital costs. They either have no screening and storage or may only have a screen prior to the next stage. The decker and storage tower do not generally improve quality or reduce costs except in cases where excess water from the papermill can be used for dilution and useful fibre recovered. However, for overall operating efficiency, the brown stock storage has many advantages. Pulp production can continue during a failure of the bleaching or paper mill section or during a planned shut-down of either unit.

I.3.5 Cleaning

Cleaning is desirable for all types of pulp and is essential for bleached grades. The small mill normally cleans the pulp before bleaching in order to avoid the cost of another decker before the paper machine. Cleaning is done at a low consistency of 0.5 to 1 per cent with centrifugal cleaners which are available in various types. Sand, grit or other undesirable constituents are removed during the process. Cleaning can be single-stage, two-stage, or three-stage. The difference between these systems lies in fibre economy, in view of the fact that rejects are continuous and contain fibre. In two-stage cleaning, rejects from the first cleaner are processed through a secondary cleaner while a three-stage system takes secondary rejects through a tertiary cleaner before final rejection. Cleaning is expensive in both capital cost and power consumption. Thus, most small mills tend to use two-stage cleaners.

I.3.6. Bleaching

Bleaching is normally carried out on a three-stage basis. However, very small mills may operate with a single-stage bleaching only. The bleaching process removes the residual non-cellulose material from the pulp since it cannot be removed by cooking without a lowering of both pulp quality and yield. In the traditional three-stage system, chlorine (in gaseous form) is mixed with the pulp, thickened to a consistency of about 6 per cent and fed to an upward-flow chlorination tower. The retention time ranges between 45 to 60 minutes. The mixture is subsequently diluted and washed over a washer. It then passes through a heater-mixer, where the temperature is raised to about $45^{o}C$. Caustic soda is added and the pulp is finally transferred to another retention tower for a period of approximately an hour and a half. This is termed the extraction stage because it removes the colouring matter of chlorinated and oxidised lignin arising from the first chlorination stage. This operation enhances the ultimate brightness while reducing the requirement of bleaching chemicals and ensuring an optimum retention of fibre strength.

After caustic extraction, the pulp passes through another washer. Its temperature is then raised to about $50^{o}C$ in a heater-mixer where hypochlorite is added. It is transferred to a retention tower (usually of a down-flow type) where it is held for a period of two to two and a half hours. Subsequently, the pulp passes through the final washer and then to the bleached pulp storage tower which, ideally, should be capable of holding a minimum of 12 hours production. The hypochlorite stage is a bleaching stage, and single-stage bleaching can be achieved by using hypochlorite alone. Fibre quality can suffer if excessive hypochlorite is used. However, the correct combination of chlorine, caustic extraction and hypochlorite provides the most effective and economic bleaching sequence for agricultural fibres. Hypochlorite may be sodium based (a blend of caustic soda and chlorine) or calcium-based (a blend of chlorine and slaked lime). Developing countries which produce limestone should preferably use calcium hypochlorite. The latter is usually less expensive and the filtrate is relatively innocuous, thus making it easier to dispose of. In the hypochlorite stage, the pulp obtains most of its whiteness. It is less of a residual lignin removal process than the preceding stages. Furthermore, the bleaching is mostly an oxidisation process which can, if exceeded, cause cellulose degradation.

For straw or bagasse pulps, the first chlorination stage usually requires 50 to 60 kg of chlorine per tonne of pulp (i.e. 5 to 6 per cent chlorine). The caustic extraction stage uses 2 to 3 per cent of caustic soda, and the calcium hypochlorite stage 3 per cent of chlorine. Lime requirements for the hypochlorite stage amount to 40 per cent of the chlorine used.

Straw pulp reaches a brightness of 72° GE or slightly above. Although this index does not match the extreme brightness of five-stage, fully bleached wood pulp (85° GE to 88° GE), straw pulp is quite suitable for the grades of paper manufactured. For example, standard newsprint is only 60° GE. Bagasse pulp can reach a brightness of 80° GE from a three-stage process. This level of brightness enables it to compete favourably with wood pulp.

The loss of yield from bleaching is about 10 per cent of the unbleached pulp processed. Thus, the overall bleached pulp yield from straw is around 32 per cent while that from bagasse is around 45 per cent of the prepared raw material. The water requirements for bleaching are considerable, and the effluent quantity correspondingly high. Each of the three washers need a minimum of 40 tonnes of water per tonne of pulp washed. The chlorination and hypochlorite stages are strongly acid, thus requiring suitable materials for the equipment and storage. The caustic extraction stage is alkaline and does not require special materials.

I.4. Pulping stages for other fibre materials

The previous sections described a typical pulping and bleaching process commonly used for the production of pulp from straw or bagasse. More details are provided for these two materials because they are widely available in most developing countries. They can be used for the manufacture of cultural papers which are the types of paper most required in these countries, apart from newsprint . However, less widely available materials may also be used by small-scale mills. These materials will require adjustments in the processes described above, depending on the type of material and that of the end product, the scale of production and the availability of investment funds. Furthermore, some adjustments can also be made in the processing of straw or bagasse. For example, some stages can be omitted at the expense of quality or efficiency. Thus, single-stage washing and bleaching are sometimes adopted by mills with a capacity of 15 tonnes per day. The paper product is in this case only suitable for scholastic use or cheap stationery. The blow-tank and high-density storage may be omitted. The latter is not practical except where vacuum type washers are employed. In many of such instances, capital cost is the dominating factor. However, if a local market exists for low quality products, small mills can still be viable. They may also improve quality, product range and capacity in order to suit market demand, provided investment funds can be made available.

Raw materials other than straw or bagasse usually require a different treatment. It is not uncommon to have two streams of pulp production in the same mill. For example, some long-fibre pulp may be produced in order to strengthen the short-fibre straw or bagasse pulp. This is often the most difficult problem faced by small mills making cultural bleached papers or unbleached wrapping grades. A brief description of the processing stages for other materials is provided below.

Cotton waste
If available at economic prices, cotton waste (i.e. in order of preference, clean cuttings, cotton linters or selected rags) is the best long-fibre source for bleached paper. It provides the strength requirements with one-third less material than would be required by softwood pulp in view of the fact that cotton fibres are longer and stronger. It also requires less chemical for cooking because cotton contains at least 80 per cent of cellulose. Consequently, a simple single-stage bleaching process is sufficient to achieve high standards of brightness. The inherent quality of the paper is also improved. The scale of production is usually small because a 20 per cent addition of cotton pulp is sufficient for all but art papers. Furthermore, cotton waste is seldom available in large quantities.

Rags should be hand-sorted in order to eliminate synthetic materials which cause problems. Cuttings and rags are chopped and dusted, using a rag chopper cycle of three to four hours (at 4 to 6 atmospheres steam pressure) with 4 to 5 per cent of caustic soda. The cooked pulp is then dumped onto a draining floor beneath the digester. After the cooking liquor has drained away, the pulp is taken by conveyor or by hand to a series of breaker washers. It is then dumped in draining compartments and left for approximately 12 hours to mature. In some small mills, the pulp is dumped directly from the digester into a large potcher-washer mounted beneath. After washing and maturing, the pulp goes to beaters for refining. Calcium hypochlorite is then added in the proportion of 1-2 per cent to achieve the desired brightness. The overall yield from good cotton waste can exceed 80 per cent. A flow diagram for cotton pulping is shown in figure IV.4.

Jute and hemp waste

The pulping process for jute and hemp waste is similar to that applied to cotton. However, the pulp is not bleachable in view of the characteristics of the raw materials, which include used gunny sacks, old rope and hessian . The yield is lower (about 60 per cent) and the pulp is used almost exclusively as the long-fibre component of wrapping grades of paper. Hemp can be used for specialty papers, such as cigarette paper. However, longer cooking with more chemical is required because this grade must be bleached.

Unbleached pulp

Unbleached pulp for wrapping grades is cooked "harder" than bleaching grades (i.e. more lignin is left in the pulp, less caustic soda is used and a higher yield is obtained). The range of wrapping papers varies from pulp for fluting paper (based on straw or bagasse cooked with 6 per cent of caustic soda and having a 70 per cent yield) to lightweight wrapping grade pulp (cooked to approximately 50 per cent yield with 8 to 10 per cent of caustic soda).

In general, the coarser the paper to be made, the higher the yield. The softer cooked grades are used for lightweight bag paper or similar products.

II. CHEMI-MECHANICAL PULPING

For the very small integrated mill (e.g., 10 to 15 tonnes per day), chemi-mechanical pulping is less expensive in capital cost, has a higher yield, consumes less chemical and requires less water. It is also a very simple, manually operated process. However, the quality of pulp produced is inferior to that obtained from the fully chemical process although it may be satisfactory for specific markets. This process applies mostly to straw,

- 53 -

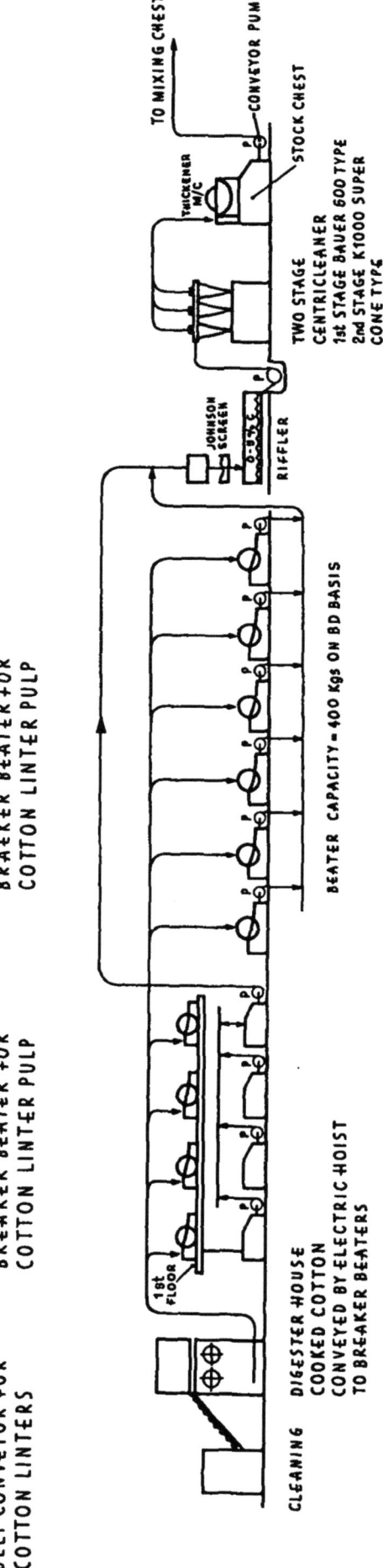

Figure IV.4

FLOW SHEET FOR COTTON PULP

bagasse or reeds and generally, but not exclusively, to unbleached grades of paper. Single-stage, hypochlorite bleaching is used in some mills for low brightness products.

Cooking is carried out batchwise in a vertical, cylindrical vessel with a rotating, bladed impeller. In both construction and operation, the vessel is similar to the hydrapulper which is used internationally for slushing waste paper or imported pulp. The chopped fibre is fed into the pulper and water is added until a consistency of around 10 per cent is obtained. Steam is injected to raise the temperature to the boiling point (100°C) because the pulper is open to atmosphere. A variable amount of caustic soda (5 to 10 per cent) is added according to the grade of pulp required. The straw is cooked and mechanically defibred by the impeller for a period of approximately one hour and then discharged into a stock chest. Hot water is normally added during the discharge in order to lower the consistency to approximately 5 per cent. This operation facilitates pumping and subsequent refining. Since the degree of de-fibring is lower by this method than required by the end product, the pulp is pumped from the dump chest to a refiner for further mechanical treatment. The refining is most effective when the pulp is still hot. Furthermore, quality is improved with an increase in the consistency of the pulp refined. However, consistencies above 6 per cent require special, more expensive pumps and refiners. The disc type of refiner is the most suitable.

After refining, the pulp passes over a washer and is then ready for the paper machine. Alternatively, it undergoes a final bleaching operation after washing if bleached grades are required. For a given brightness standard, pulp obtained by this process requires more bleaching chemical than is needed for fully cooked pulp. For this reason, chemi-mechanical pulp is more suitable for unbleached or semi-bleached grades. A single pulper may supply a mill producing 10 to 15 tonnes of paper per day. The overall yield on straw is 45 to 60 per cent, depending on the type of output. This pulping process is, under suitable conditions, particularly attractive for investors with limited financial means.

The advantages of chemi-mechanical pulping (higher yield, lower chemical consumption and capital costs) are also recognised by large-scale producers of unbleached grades. It is possible to increase yield and capacity from an existing mill by installing a hot-stock refiner after the blow-tank. This will also result in lower chemical costs per tonne of pulp produced. For new installations, the capital cost economy has led to relatively sophisticated arrangements, incorporating pre-washing and dewatering, and a battery of mechanically fed pulpers.

At the lowest quality level and smallest scale of production, lime can be used instead of caustic soda. However, this will produce only the cheapest grades of straw paper for wrappings or inferior fluting. Figure IV.5 provides a flow diagram of chemi-mechanical straw pulping.

III. SEMI-CHEMICAL PULPING

High yield pulp, most commonly used for corrugating medium paper, can also be made with a blend of sodium sulphate (Na_2SO_3) and sodium carbonate (Na_2CO_3) instead of caustic soda. This process, termed the neutral sulphite system, is most commonly used in developed countries for wood-based products. However, it is not recommended for the small mill in developing areas for the following reasons: chemical preparation is more difficult; chemical recovery is not possible; the pulp is corrosive unless very thoroughly washed; and the yield or quality advantages are relatively insignificant when straw or bagasse is the raw material.

IV. CHEMICAL RECOVERY

Assuming a 40 per cent yield and a 10 per cent caustic soda consumption, the production of 1 tonne of bleachable grade stock from straw results in an effluent comprising 20 tonnes or more of brown liquor. The latter contains 1,500 kg of organic material removed from the straw and 250 kg of inorganic chemicals. The organic material is combustible if the water content can be sufficiently reduced. In the traditional chemical recovery, the brown liquor is evaporated to around 60 per cent of solids. At this density, it is viscous (resembling heavy fuel oil) and can burn. If caustic soda is the cooking chemical, the latter is reduced to smelt during combustion and becomes soda ash. If treated with lime, the soda ash reverts to caustic soda again and can be re-used. The heat produced during combustion arises from the organic material. When the pulp yield is 50 per cent or less, the heat is theoretically sufficient for the evaporation and recausticising operations and for the production of steam for the digestion stage. Thus, in addition to chemical recovery, the operation produces enough heat for most of the plant requirements. Furthermore, the lime used can be recovered by re-burning the sludge.

Figures IV.6 to IV.8 provide the flow sheets for a roaster smelter, a causticising plant and a Sulzer-Escher-Wyss recovery system. Figure IV.9 shows a picture of a Sulzer chemical recovery system.

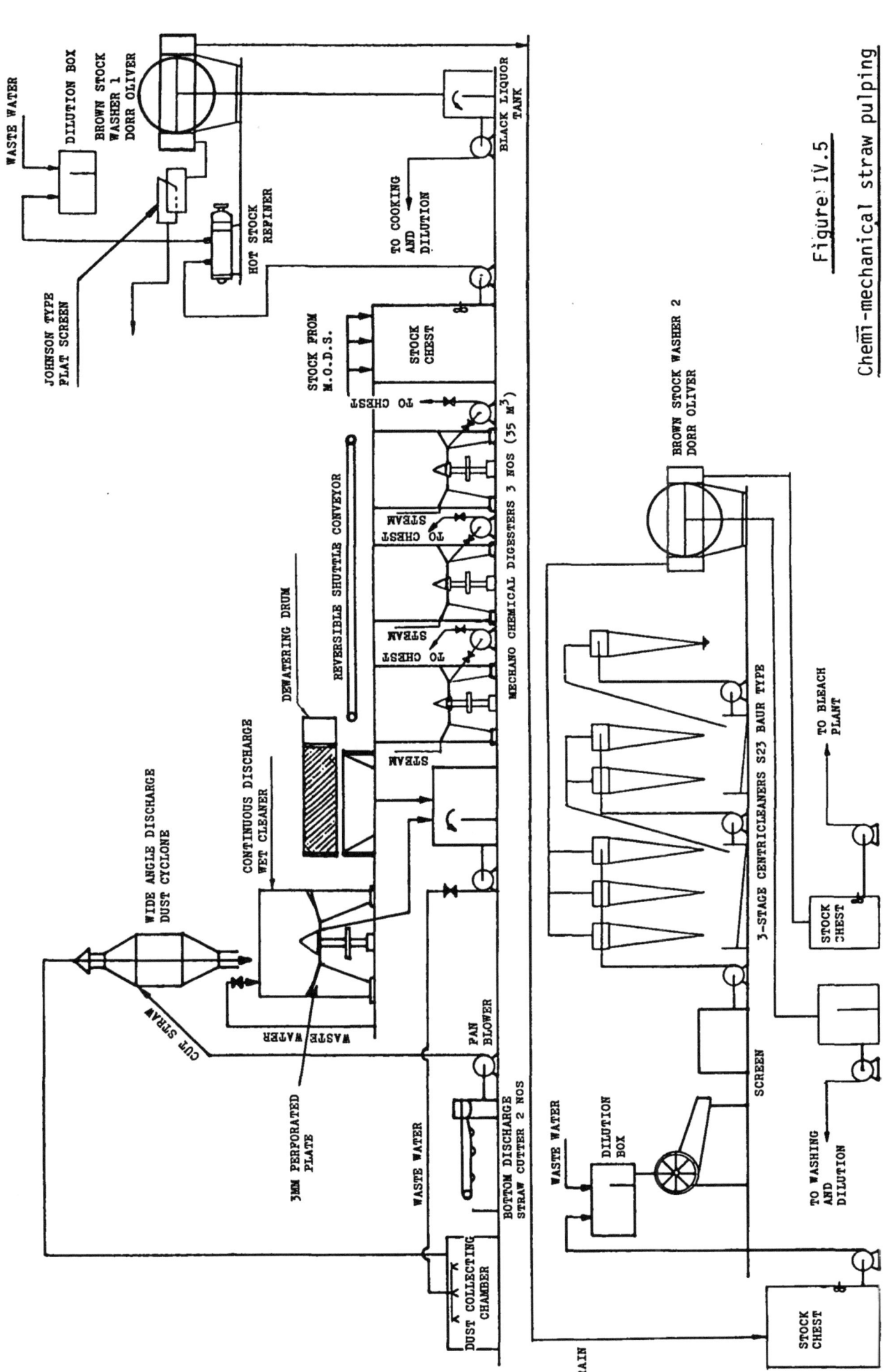

Figure IV.5

Chemi-mechanical straw pulping

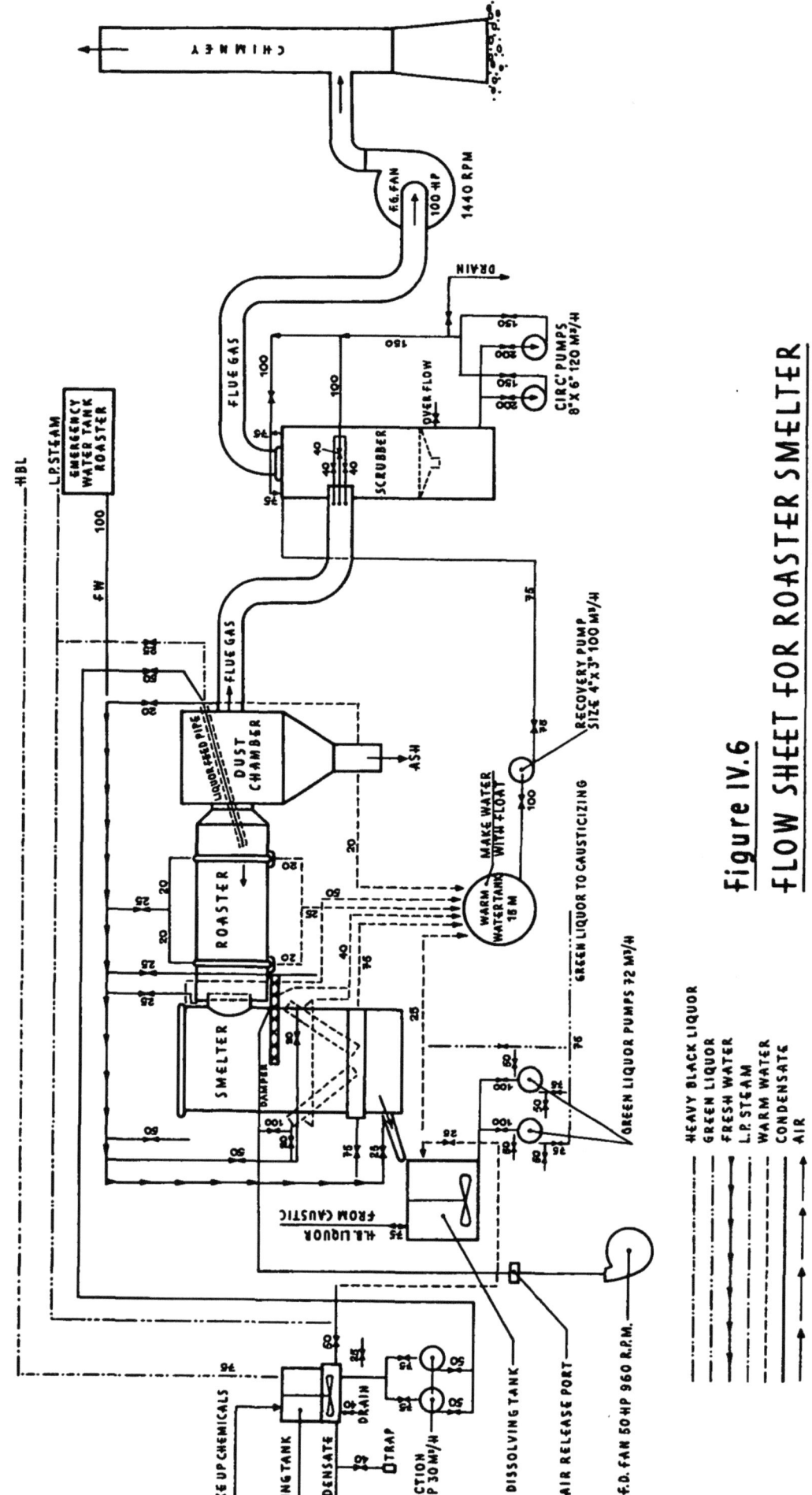

Figure IV.6
FLOW SHEET FOR ROASTER SMELTER

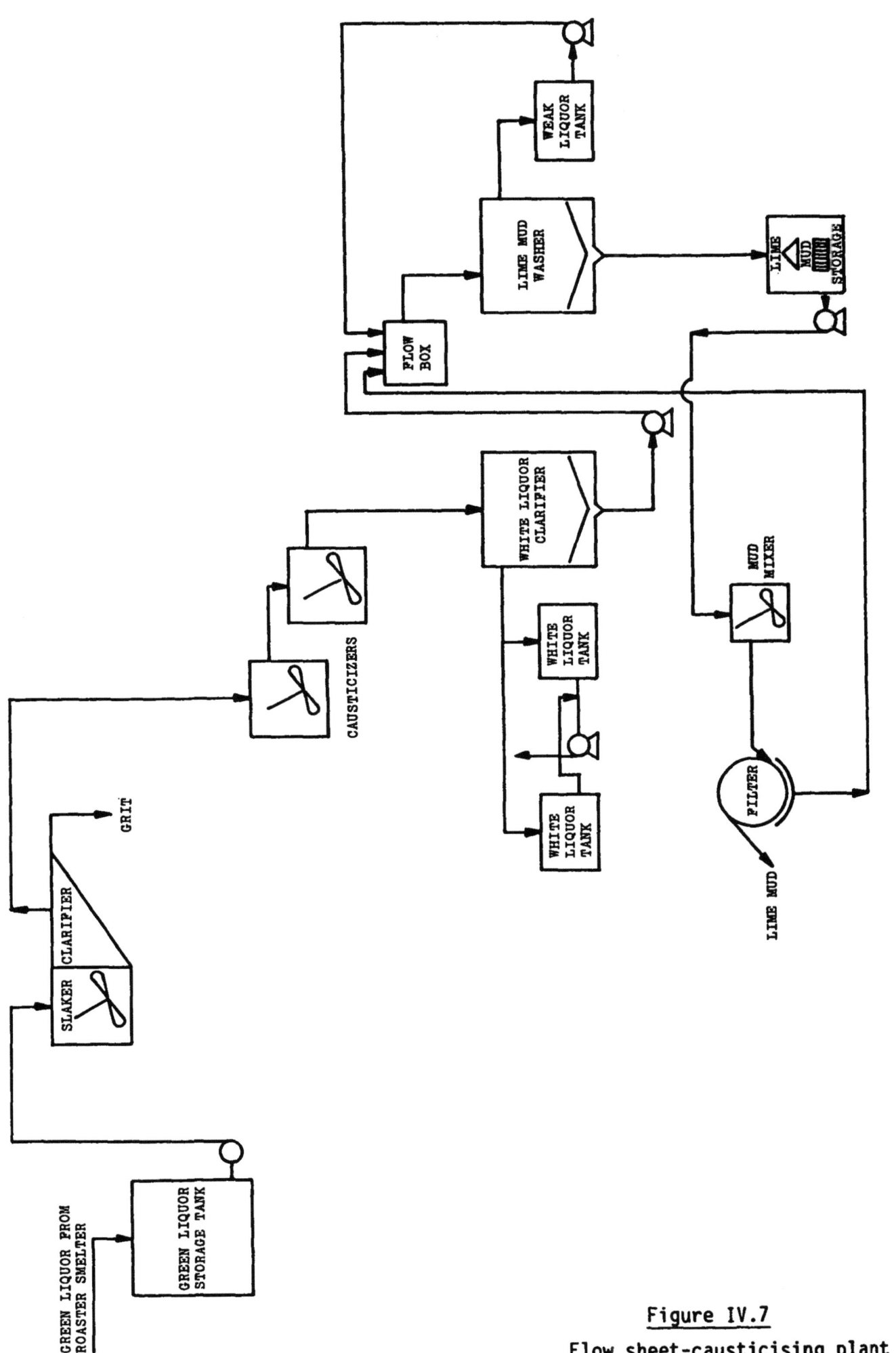

Figure IV.7

Flow sheet-causticising plant

- 59 -

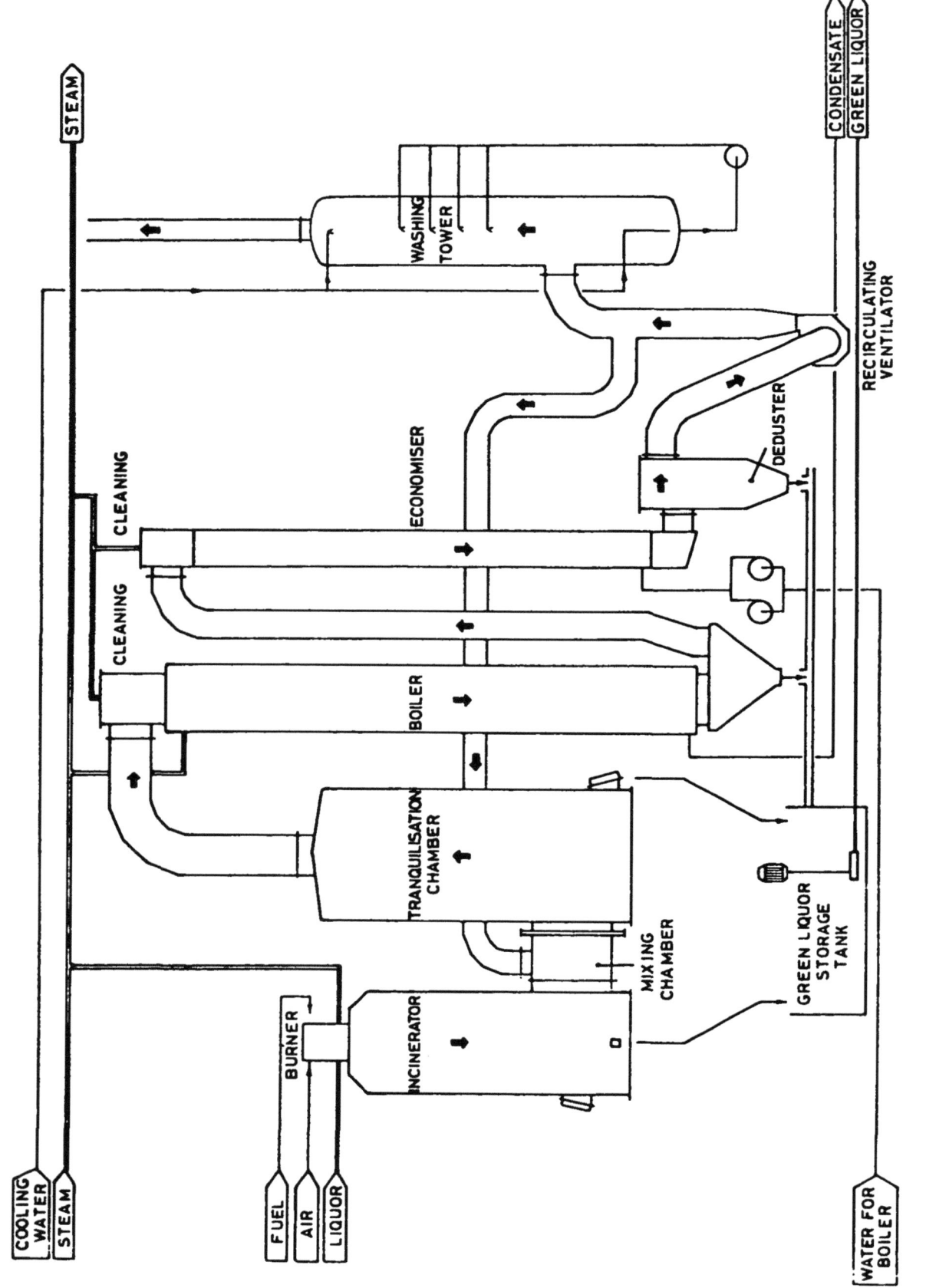

Figure IV.8

SULZER-ESCHER-WYSS RECOVERY SYSTEM

- 60 -

Figure IV.9

Sulzer chemical recovery system

The heat and chemical recovery process just described can also be achieved in large pulping mills. These mills (200 tonnes per day capacity or above) do recover around 85 per cent of the chemicals and 90 per cent of the lime while supplying the heat requirements for the entire pulp mill and contributing to the energy requirements of the paper machine. The burden of effluent disposal is also eliminated. However, the recovery boiler is a highly technical and sophisticated unit, and is therefore very expensive. It is not an economic proposition for capacities below 100 tonnes per day of pulp. Thus, the full benefits of chemical recovery may not be realised for mills of lesser capacity. In this respect, therefore, scale does have an undeniable advantage.

However, the greatest economic benefit lies in the chemical aspects of recovery. This is particularly the case for small mills located in countries where caustic soda is not produced economically or may have to be imported. It is possible to recover up to 75 per cent of the chemical used in mills with a capacity as small as 20 tonnes per day.

For mills with a capacity 50 or more tonnes per day, some heat recovery is practical and may be sufficient to supply the steam required to evaporate the brown liquor to a point at which the residue can be burnt. In this case, the residue is burnt not in a boiler but in a kiln or fluid-bed burner. Waste-heat recovery can be incorporated.

For mills producing 20 tonnes per day of pulp, 70 per cent chemical recovery is still feasible. However, the expense of heat recovery cannot be justified and the economy in chemical recovery is offset, to some extent, by the necessity to provide steam for evaporation. A "roaster", or simple combustion chamber, is used to burn the residue of the evaporated brown liquor. Evaporation is not carried beyond the degree sufficient to support combustion. Thus, three-stage evaporators are sufficient (as against five-stage evaporators for recovery with steam generation) and capital outlay is therefore reduced.

Chemical recovery from straw cooking is also more difficult than from wood because of the high silica content of straw. It may be necessary to accept a lower level of evaporation (and heat generation) to avoid tube-scaling and it is not possible to recover the causticising lime. Moreover, the small recovery plant can seldom justify a lime kiln suitable for recovery. Disposal of lime mud can be a problem for large straw-based mills.

To a lesser extent, bagasse mills may also have a silica problem but at least 50 per cent of the lime should be recoverable if the scale of production is sufficiently large to justify investment in a kiln.

For pulp mills below a capacity of 20 tonnes per day, chemical recovery is unlikely to be feasible unless the cost of caustic soda is high. It was indicated in a recent study by an Indian consultant that if the price of caustic soda were to increase substantially, chemical recovery could become feasible at a capacity of 15 tonnes per day.

While the traditional chemical recovery process may not be feasible for small pulp mills, a new patented recovery process may offer a suitable alternative. The Copeland process, which is still being developed, is used for the treatment of pulping liquors with a sodium base. It is based on the thermal oxidation (at approximately $700^{o}C$) of the organic matter contained in a pre-concentrated solution of the waste in a fluidised bed reactor. The Copeland process operates autogenously (without the introduction of supplementary fuel) at concentrations of total solids considerably lower than those required for other types of recovery furnace (30 to 35 per cent as against 55 to 65 per cent). It is particularly suited for small mills which use raw materials with a high silica content that tends to scale the traditional recovery plant equipment. Its advantages over the roaster smelter system include better thermal efficiency, a higher chemical recovery rate, almost continuous operation throughout the year, and no effluent disposal problem. For further information on this process, the reader may write to: Hangal Paper Consultants Private Limited, A-7 New Friends Colony, New Delhi 110065 (India).

V.   EFFLUENT DISPOSAL

Small mills producing 50 tonnes of pulp per day or less cannot justify investment in the equipment required for the processing of the black liquor. Consequently, disposal of the latter will be necessary. With a yield of 65 per cent and a chemical consumption of 6 per cent, black liquor containing approximately 530 kg of organic material and 90 kg of inorganics is produced per tonne of pulp. The major burden for effluent disposal mostly relates to the organic content of the liquor.

It is possible to reduce the organic content of the liquor by approximately 80 per cent and to recover combustible gas for the mill operation. This can be done through the fermentation of the organic material in an anaerobic digestion process. The latter was used in a pilot plant, and current estimates show that 30 per cent of the total fuel requirements of a mill can be produced through this process in the form of methane. For a lower yield of pulp the gas generation potential is greater.

Investments in an anaerobic digestion unit are relatively low. Furthermore, the unit is simple to operate once the parameters have been evaluated, and probably offers the most promising field of effluent treatment for the small pulp mill. However, the chemical content is not recovered, and in areas where strict legislation on effluent disposal obtains, further anaerobic, biological treatment may be necessary. This is seldom the case in rural areas in developing countries where industry and employment are the predominant concern of governments. It is, however, inevitable in the long term that adequate effluent treatment be enforced. The small mill can achieve this objective economically by applying one of the processes suggested above. It is strongly recommended that the necessary pollution control equipment be incorporated from the start whenever investment funds are available.

Bleached effluents contain less organic material and their disposal is therefore facilitated. The problems of effluent disposal are easier to solve for the small mill. They seldom require more than pH control and clarification by settlement. In developing areas where water is scarce, the effluent is often used by local farmers for irrigation. Given a sufficiently large area of dispersion, this method could be fairly beneficial. The liquid flows in open drains over distances of 3 to 4 km thus achieving a degree of clarification and oxidisation which greatly improves its quality.

VI. STOCK PREPARATION

Pulping reduces the fibre source to a suspension of fibres in a slurry of water by physically tearing them apart (mechanical pulping) or by dissolving the natural bond (chemical pulping), or by a combination of both processes. Subsequently, the pulp must undergo a further mechanical treatment prior to processing in the paper machine. Fibres must be separated one from another, brought to a uniform size which is dependent on the paper grade involved, and subjected to a final processing in order to obtain the desirable paper quality and characteristics.

Where substantial quantities of the natural bonding material (lignin) remain in the pulp (as is the case after mechanical or semi-mechanical pulping), the lining enclosing the fibres limits the work which can be done.

On the other hand, the fibres are relatively free of lignin after chemical pulping. They can be bruised, flattened and fibrillated, thus increasing the surface area and hence the paper bonding mechanism. This also improves the nature of the bonds between the fibre surfaces, and strength characteristics can be developed. The mechanical treatment (termed "beating" or "refining") must be carried out in the presence of water.

Beating is normally a batch process while refining is usually a continuous process although it may also be carried out on a batch basis. There are several types of beater and many designs of refiner. The processing of the fibres is, however, essentially the same for all types of equipment: the pulp, as a slurry of fibres and water, is made to pass between metal bars, or plates, exhibiting a barred pattern; pressure is applied between the bars; the work done on the fibres being a function of the pressure applied between the bars or plates and the number of bar edges to which the fibre is subjected in a given time. The work done on the fibres is not a measure of the total power consumed by the beater or refiner: there is always more water than fibre in the slurry and energy is partly absorbed in turbulence and pumping.

If waste paper or imported dried pulp is used for the furnish, it has to be reduced to a pulp before beating or refining. The unit most commonly used for this purpose is a Hydrapulper. There are various types of pulpers with a wide range of capacities. Prior to the hydrapulper, a "breaker", or a Watford pulper, was used, the latter being a high-density pulping machine. The breaker and the Watford pulper still have relevance for small mills. Each has characteristics which make it suitable for the particular raw material processed in these mills.

VI.1. The hydrapulper

Hydrapulpers are open-topped, cylindrical vessels in which an impeller revolves. The impeller has projecting blades or "whangers" on the surface which create a maximum turbulence within the vessel. The latter may also have bars welded down its sides for the same purpose. The vessel, or tub as it is commonly termed, has a false bottom while the inner bottom has drilled holes which permit the removal of pulp or waste paper from the tub once they have been sufficiently disintegrated. Pulpers of this type can work on a batch or continuous basis. Originally, they were designed for batch operation and the perforations were large enough (up to 2.5 cm in diameter) to allow for quick emptying of the tub.

The pulping process is most efficiently carried out at around 8 per cent consistency. However, emptying is slow and pumping difficult at such a level. Thus, it is normal to add water during the emptying period in order to reduce the consistency. When used for waste paper processing, the pulper can also perform the pulp cleaning operation as follows. As the paper disintegrates, a vortex forms. A rope, fastened to the side of the pulper, is allowed to trail in the stock near the vortex to catch strings, wires, rags, etc. For batch waste paper processing, the pulper retains contraries. For mixed, unsorted waste, the proportion of such material (up to 13 per cent) is such that it is necessary to shut the pulper at frequent intervals and manually remove the rubbish. Figures IV.10 and IV.11 show respectively a cross section and a picture of a pulper.

Junk removers are devices which are built onto the side of the pulper and continuously screen a proportion of the pulp in order to remove contraries. It is therefore logical to adopt a continuous pulping system and to reduce the size of the perforations, thus removing more contraries during pulping. This method will also yield a greater degree of disintegration. However, the power requirements increase as more work is done.

The consistency in a continuous pulping system must be constant. Originally, it was maintained at approximately 2 per cent for waste paper in order to allow the direct use of centrifugal cleaners after pulping. The thick stock cleaner was subsequently developed, and a 4 to 5 per cent consistency is now common. A more recent development used for waste paper, makes use of the previous large perforations while adding a secondary, small volume pulper (called a de-fibrer) after the main pulper. This system has built-in screens for the controlled removal of small rejects and devices for continuously rejecting contraries which do not sink (e.g. foamed plastic or cork particles). However, it is not feasible to operate pulpers in this manner on a scale of less than 100 tonnes per day. Small mills are therefore not likely to adopt this technique.

Pulpers are usually vertical but can also be horizontal. The capacity range is wide, from around 10 to 300 tonnes per day. For the same size unit, capacity at continuous operation is greater than for a batch pulping operation because there are no filling and emptying times. However, batch pulping may still be preferred for the disintegration of imported pulp because there are no contraries to remove, a degree of consistency control is inherent in the process if the pulp is weighed or even counted as bales before filling, and

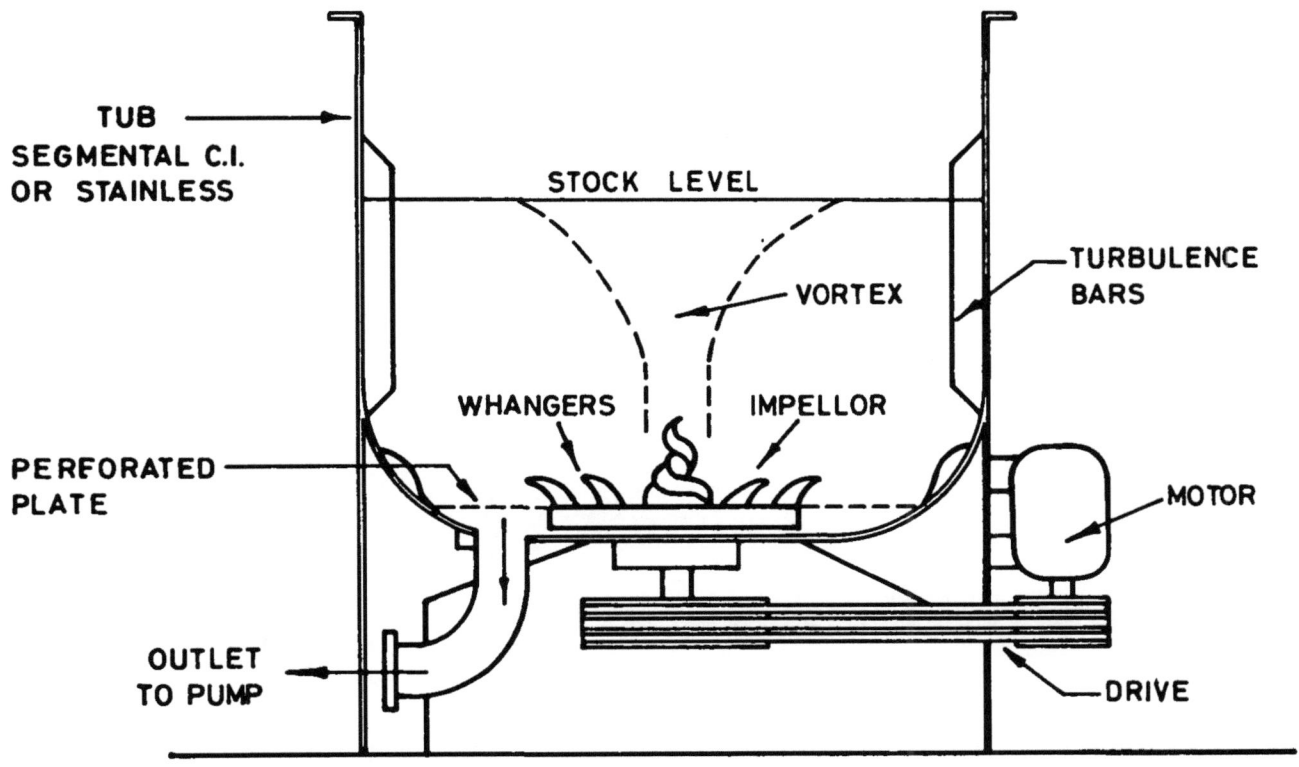

Figure IV.10

SIMPLE PULPER, BATCH OR

CONTINUOUS PULPING.

Figure IV.11
Typical pulper

there is a greater assurance of uniform disintegration. Batch pulping also permits the separate pulping of different materials for later blending in controlled proportions.

The modern pulper is normally fed by a mechanical conveyor. Small mills in developing countries may, however, avoid the acquisition of this equipment and feed in the material by hand. Small, simple pulpers can be built locally, even in countries where engineering resources are limited. It is also possible to import the impeller and drive and make the tub locally.

VI.2. Breakers

The breaker is probably the earliest form of pulper, and is still used in small mills. It comprises a heavy, cast-iron roll with its periphery slotted for the mounting of iron, bronze or stainless steel bars held in place by wooden packing strips. It is mounted on one side of an open tub which is divided by a mid-feather leaving both ends clear. The floor of the tub is sloped from an elevation, or backfall, behind the rotor, around the tub and back to the front of the roll. There is a narrow gap between the rotor and backfall.

In practice, the breaker is filled with water to a given level, the roll is revolved and pulp or waste is added. The projecting blades of the rotor create turbulence which breaks up the pulp or waste and lifts the suspension over the backfall. Subsequently, it passes around the tub and back again to the rotor. This is a batch process: when the pulp is sufficiently disintegrated, a plug at the bottom of the tub is lifted and the stock falls into a chest for further treatment.

Breakers can also be used for cotton, jute or rag pulps. In fact, it is probably the best unit for such a purpose because of the stringy nature of these materials. A washing drum can be fitted in the tub, opposite to the rotor, for pulp washing. The breaker washer can also be used as a bleaching unit for simple, single-stage bleaching. The tub can be made of sectional cast iron or locally built in concrete. It can be lined with tiles for high quality pulp. The rotors are simple and can also be made in a local foundry.

For waste paper or imported pulp disintegration, the breaker has largely been superseded by the hydrapulper. However, it is still popular for the processing of cotton or jute pulp. It is hand fed, simple and inexpensive in terms of capital. However, energy consumption per unit weight of pulp produced is relatively high. It operates at around 5 to 6 per cent consistency. Figure IV.12 shows the cross and longitudinal sections of a washer breaker.

VI.3. Watford pulper

This unit is essentially a cast iron, conical casing in which a shaft equipped with shaped, cast iron paddles revolves. The paddles have a screwing action on the material and move it from entry at the large end of the unit to removal at the small end. The pulper operates at a high consistency (around 35 per cent). The pulped material, of a crumb-like character, falls into a collecting basket or trolley for removal. The pulping action is close to a kneading process. Steam or chemicals can be added. The machine is versatile, and is capable of pulping waste of the most stubborn kind, including old paper sacks and wet-strength papers. Papermakers appreciate its effect of dispersing asphalt or bitumen when steam is used. It does not clean waste but, since it is usually hand fed sheet by sheet, contrary removal does not pose a problem. It has recently been used in developing countries for the de-inking of newsprint. Steam and chemicals are added as part of the process. It is low in energy consumption because the amount of water is small. However, it has a small capacity (less than 1 tonne per hour). For this reason and because of high labour requirements, it has been superseded in developed countries by the now conventional pulper. These reasons should not, however, hinder the use of Watford pulpers by small mills in developing countries given the small capacity of these mills and the low wage level.

Local manufacture of Watford pulpers should not be difficult in countries where engineering and foundry works are moderately developed. However, the first cost for such requirements as patterns would probably make production uneconomic unless a reasonable number can be sold. These pulpers can be fairly useful to small mills since they are able to recover, in particular, valuable long-fibre pulp from materials such as cement sacks, which are difficult to pulp satisfactorily by other means.

VI.4. Beaters

The beater (see figure IV.13) is similar in appearance to the breaker. Its rotor is barred in similar manner, but has narrower blades made of bronze or stainless steel. The roll is mounted on a pivoting arm, so that it can be raised or lowered in order to adjust the pressure between it and one or more barred bed plates located beneath it. The roll revolves at a peripheral speed of around 600 metres/minute, and absorbs 80 to 100 HP. The tub is normally

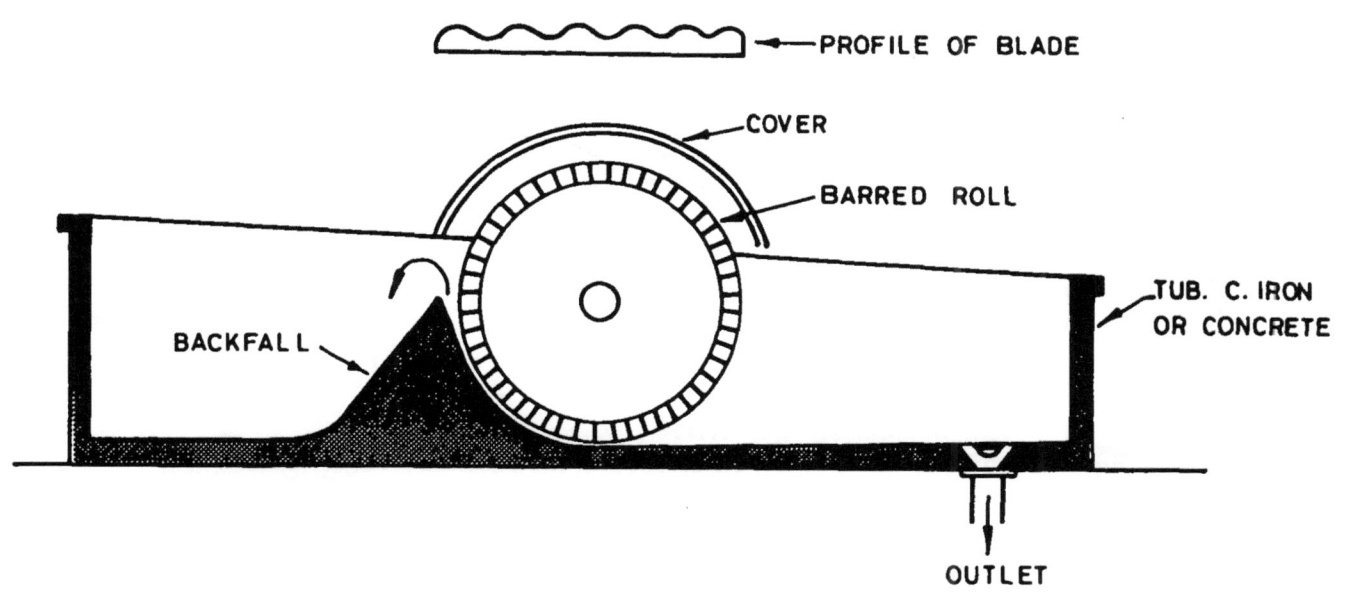

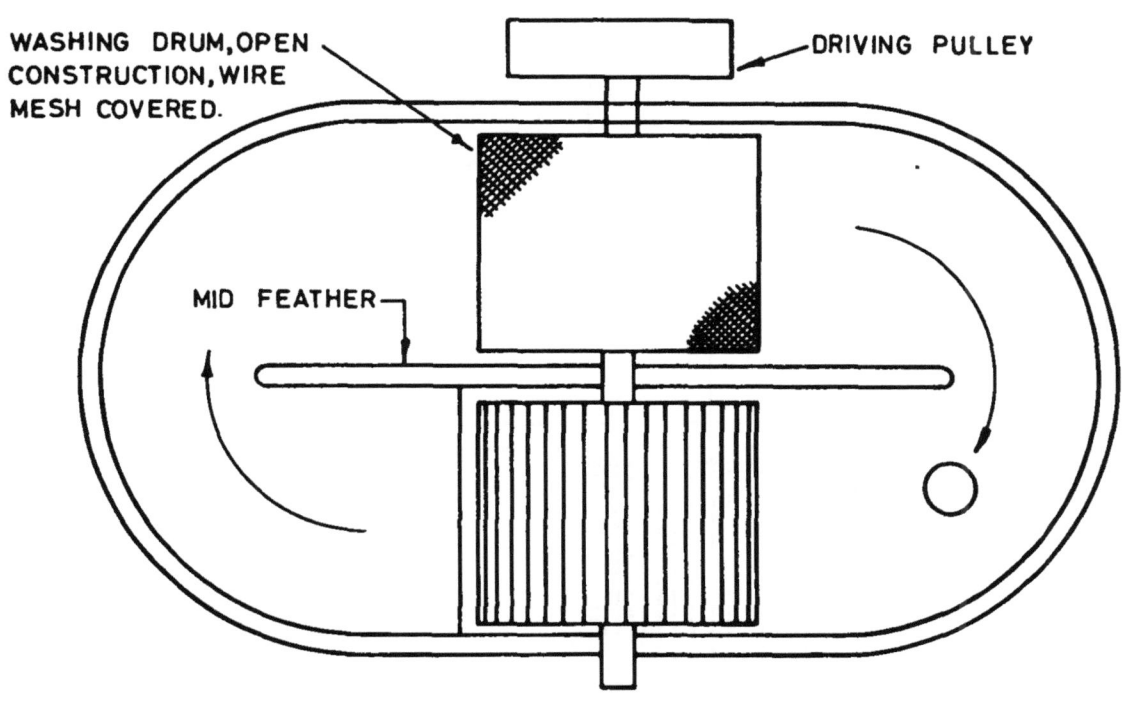

Figure IV.12

WASHER BREAKER

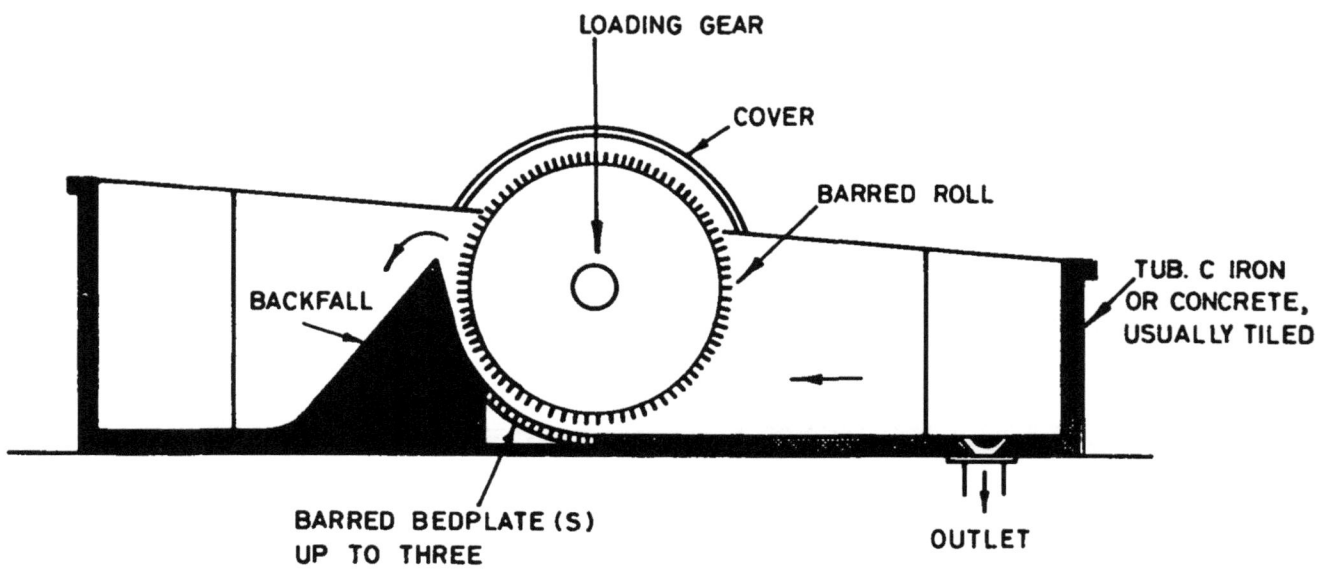

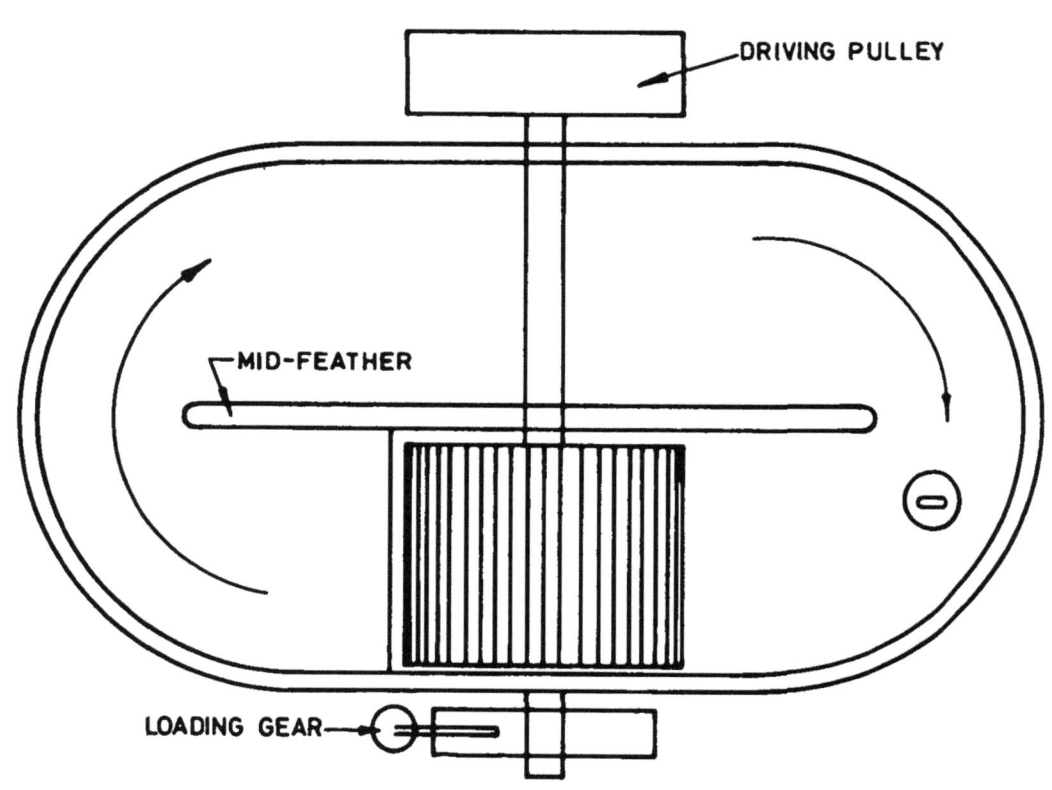

Figure IV.13

BEATER

lined with tiles, and a wooden cover is fitted above the roll to prevent splashing. The beater blades are usually 10 mm wide, but can be smaller or wider, depending on operating conditions. Narrower blades tend to cut and shorten the fibres while wider blades have a more brushing and bruising effect. For specialty papers which require intensive beating without cutting (e.g. grease-proof papers), the beater blades can be made of rock basalt and be much wider.

The operation is carried out on a batch basis from previously pulped material pumped from the breaker or pulper, or directly from the pulp mill. The beater was once the universal machine. It has since been superseded in developed countries by refiners, except for the production of low volumes of specialty papers where the desired characteristics can be obtained only by the beater. For small mills in developing countries, it is used in cases where the raw materials include textile or similar fibres (e.g. cotton, jute, hemp) for which there is no effective alternative to the beaters.

VI.5. De-flaking

In the recycling of waste paper, particularly low-grade, predominantly mechanical pulp wastes, fibre treatment is not necessary since it has already been carried out in the original production of the waste paper. Over-refining may impair the quality of the product, retard machine operation and waste power. Some treatment is however necessary in order to complete the disintegration process. The de-flaker was developed in order to carry out this operation.

De-flaker

There are many varieties of de-flakers including the conical, plate-type, barred, perforated, fixed or adjustable-gap de-flakers. However, the fundamental working principle is common to all of these: the rotating and stationary elements do not touch and there is no applied pressure between them. The power consumption is low and the fibres are not unnecessarily cut or "wetted", thus reducing the drainage rate. The major purpose of de-flakers is, therefore, the treatment of waste papers for recycling. They are very useful and inexpensive tools for small mills processing waste paper because they are available over a wide capacity range. The de-flaker is also used in some mills as a preliminary step to disc refining of imported pulp. The space between the refiner plates and that between the grooved pattern channels is small and apt to clog if the entering pulp is not sufficiently disintegrated.

VI.6. Refining

Refining is the process of bringing the pulp to the condition required for the paper product. It complements beating, and was first introduced in order to subject the pulp to a final treatment immediately before processing by the paper machine, and after beating. Refining is carried out under the control of the paper machine operator rather than that of the beater room operators. The refiner is usually used on a continuous basis, although it can also be used on a batch basis by recycling in conjunction with a chest. It requires less space and is cheaper than a beater in terms of cost per unit weight of pulp processed. It is also capable of applying much more power in a single unit, exceeding 1,000 HP for some applications as against the beater limit of 100 HP. For these reasons, it has gradually replaced the beater for all but special applications (e.g. cotton-based furnishes). The refining, or beating work is usually defined in terms of power per unit weight processed. It varies considerably according to the nature of the pulp and the required paper characteristics. For example, cotton-based pulps can absorb and generally require refining in the order of 100 HP/tonne/day. This is one reason why the beater remains the best tool because single refiners seldom contribute more than 10 HP/tonne/day and a series of refiners are needed for higher levels. With a single beater, the cycle period is a function of the raw material used and of the required end product. Softwood pulps take around 20 HP/tonne/day for writing and printing grades and a longer cycle for maximum strength papers (e.g. sack-kraft) or for papers which must be thin and strong (e.g. carbonising papers which require 30/35 HP/tonne/day).

Mechanical pulps cannot accept high refining. The limit is close to 7 HP/tonne/day. This level, or a slightly higher one, applies also to straw-based pulp for writing or printing grades. Bagasse requires very little refining (around 3 HP/tonne/day). This is a distinct advantage in terms of power economy. Bleached hardwood pulps require up to 12 HP/tonne/day for writing and printing grades.

In case the refiner is used to produce pulp instead of treating the latter, the power requirements can be much higher. Thermo-mechanical or chemi-mechanical pulp refiners are often very large units which may absorb 1,800 to 2,000 KWH/tonne. However, as previously indicated, these units are unlikely to be used by small mills which may, exceptionally, use relatively small refiners.

Immediately before the paper machine, "trimming" refiners are used for final control of the stock. Power requirements are low (2 to 3 HP/tonne/day) because this is merely a stock adjustment operation.

The effect of beating, or refining, is to develop strength or fineness. For any given fibre, the work done can be quantified by measuring the drainage rate, or "freeness" of the pulp, because the pulp becomes less "free" (i.e. it drains more slowly) with increased refining. Drainage characteristics can be measured by a "freeness" tester which is calibrated to give a numerical value in relation to one or two accepted standards: the Schopper-Riegler standard which is generally used for well-refined or beaten stocks, and the Canadian freeness standard which is more common in the newsprint and mass production grades. The skill of stock preparation consists in controlling the process in order to achieve the desired effect with minimum refining. This will reduce power consumption while increasing the paper machine output because drainage rates, and the press and drying performances can be affected by over-refining.

For small mills, the two most important characteristics of paper which are influenced by beating or refining are "burst" and "tear". Increased refining for a given pulp improves burst but lessens tear resistance. Thus, the preparation treatment should be controlled in order to achieve the optimum balance between these opposing characteristics. Refining also influences other characteristics (e.g. permeability, stretch or dimensional stability).

Refiners operate over a very wide capacity range, some of them incorporating special design features. In broad terms, however, refiners can be divided into two main classifications: conical refiners and disc refiners. These are briefly described below.

Conical refiners

The original conical refiner (see top drawing of figure IV.14) was invented towards the end of the last century and was given the name Jordan, after its inventor. It was the first continuous refining unit although it was probably originally used in conjunction with batch processing. It comprises a conical tapering shell, in which a rotor with a matching taper revolves. Both rotor and shell are made of cast iron and are fitted, like the beater, with metal bars held in position by wooden packings. The rotor can be moved along the axis of the shell in order to exert a pressure between the bars and to

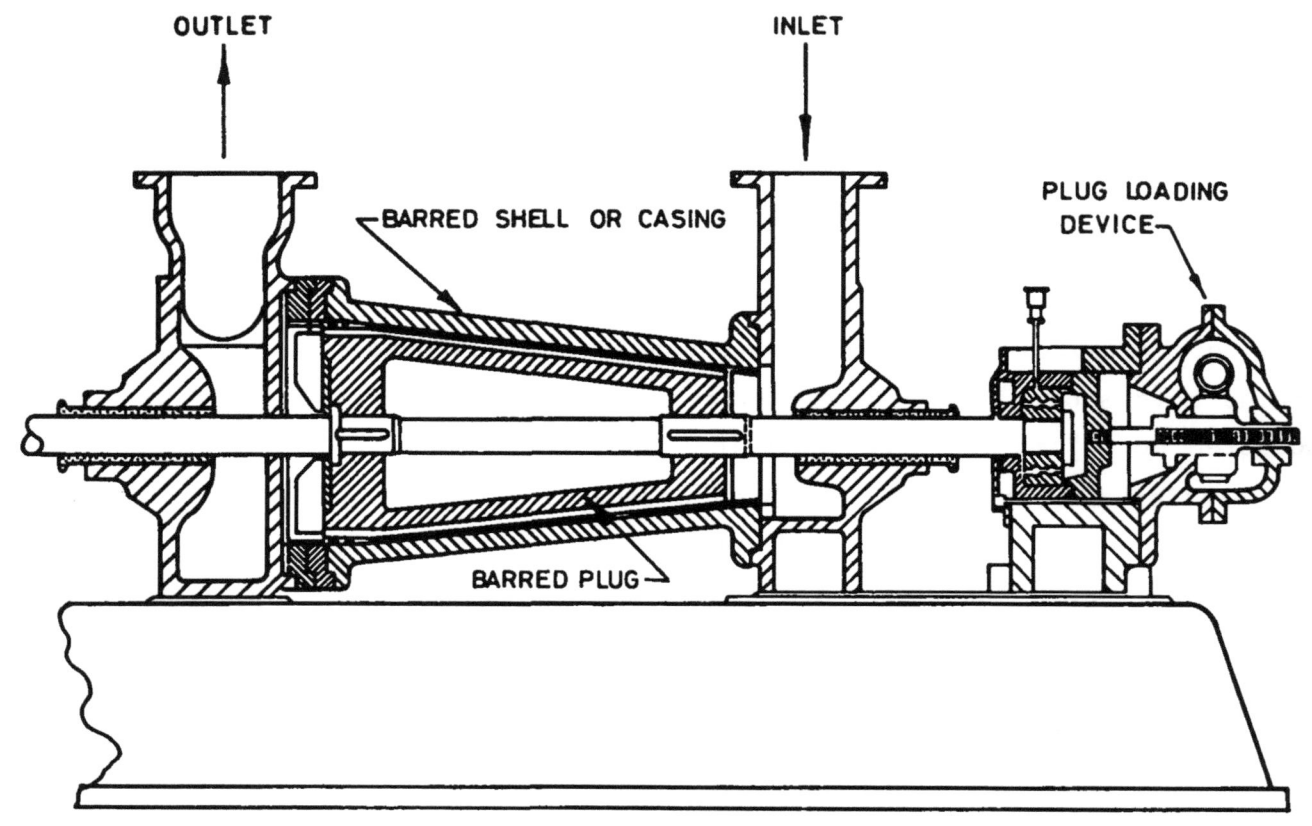

OUTLET

INLET

PLUG LOADING DEVICE—

—BARRED SHELL OR CASING

BARRED PLUG—

CONICAL REFINER

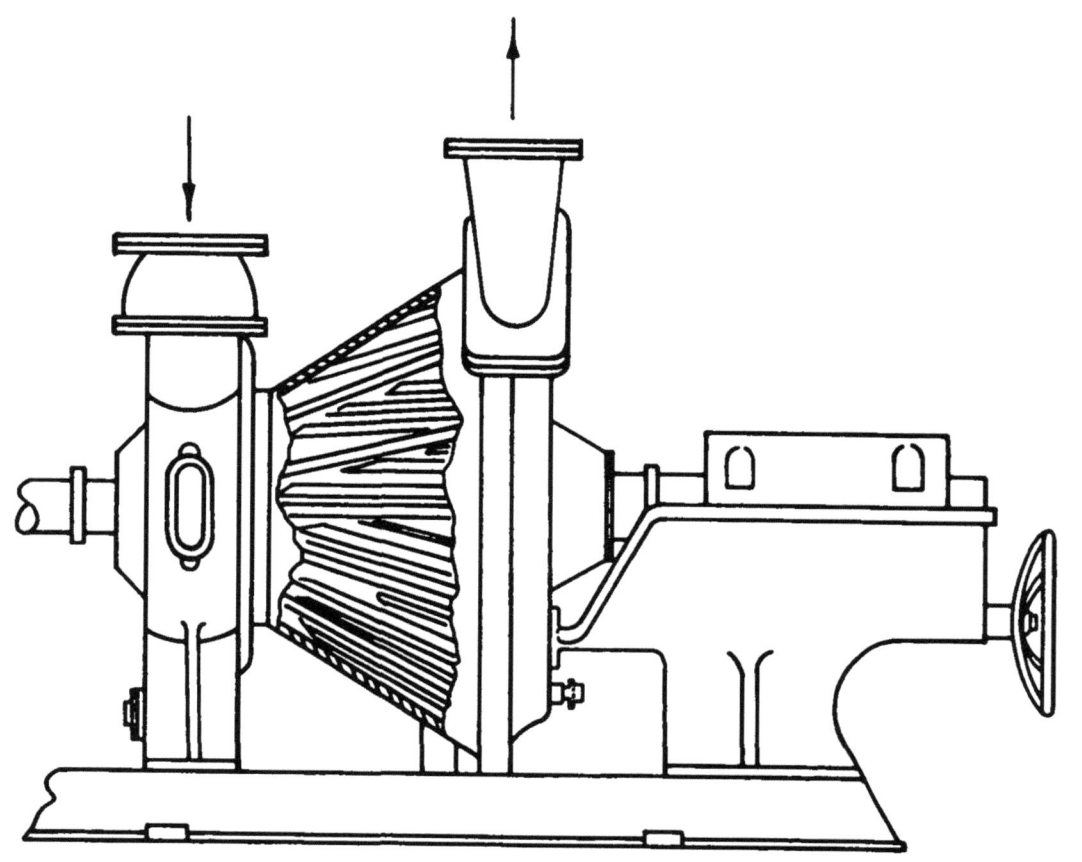

Figure IV.14

WIDE ANGLE CONICAL REFINER

compensate for wear. The rotor bars are usually axial and spaced at equal intervals around the surface. Thus, short bars must be inserted, at the large end in between the longer bars extending the whole plug length. The shell bars are shorter and usually arranged in a herring bone pattern. Stock enters at the small end and leaves at the large end. The rotor has a peripheral speed of approximately 900 m (3,000 ft.) per minute at the large end.

The Jordan-type conical refiner is still used extensively under a wide range of designs and capacities. Modern refiners may use solid cast rotors and barred stators, or may have pre-cut and assembled replaceable bars and packings. For small mills in developing countries, replaceable bars and fillings are an advantage because they cost less as spares and can often be manufactured locally. The Jordan-type conical refiner is fundamentally a refiner of low power intensity: it seldom imparts more than 5 HP/tonne/day in a single pass. More power can be applied according to the motor fitted, but such an increase may result in fibre cutting. Although bars of varying width or material do have an effect on output, the conical refiner is generally regarded as a machine with cutting characteristics. It is commonly used for final trimming because the power application is small, while the volumetric capacity is large, so that few machines are required.

The wide-angle conical refiner (see bottom drawing of figure IV.14) is similar to the Jordan-type machine but has a wider cone angle. The cutting effect is lessened while the brushing or fibrillation effect is increased. It is a refiner of medium power intensity. A maximum of 10 HP/tonne/day can be applied without excessive shortening of the fibres. It has also been claimed that some curl or equivalent is imparted to the pulp, thus improving fibre bonding. It is a versatile refiner, growing in popularity recently and available over a wide capacity range.

Disc refiners

The disc refiner consists of a circular, patterned disc rotating against a similar stationary disc. The driving shaft is normally horizontal, so that the disc faces are vertical. Stock is introduced into the centre at 5 to 6 per cent consistency and leaves at the periphery. Single-disc refiners have only one set of opposing plates while double-disc refiners have one double-sided rotating disc and two stationary discs, one on each side of the rotating member. Pressure between the discs is provided by mechanical or hydraulic means. The patterned elements of rotating or stationary discs are usually made of cast stainless steel or iron alloy. They are segmentally bolted on to the disc body and may be replaced after wear. Double-disc refiners (see figure IV.15) can usually be adjusted so as to give series or parallel flow refining within the same unit. In series flow refining, the stock must pass from one set of discs through the other, doubling the power input per unit of pulp processed. In the parallel arrangement, the pulp flow is divided into two, each half flowing through one of the two sets of plates, thus doubling the volume. The plate patterns available are numerous, the selection being made according to the type of pulp and product involved. The disc refiner is essentially power-intensive. However, since the area of contact is relatively large and holds more cutting edges than a conical refiner, the pressure on the fibres is relatively low, thus accentuating brushing or fibrillation and reducing cutting. The disc refiner can apply an energy of 10 HP/tonne/day or less per pair of plates. It has increased in popularity in cases where work of this order is necessary. Units with a capacity of 3,000 HP or less are available. However, the lower limit for effective performance is approximately 200 HP, although smaller units can be obtained.

It is important that pulp supply to a disc refiner be well disintegrated because the pattern channels are relatively small and are thus liable to become blocked if the pulp is lumpy. Although the disc refiner has advantages for fibre treatment, its effective utilisation in small mills is limited to plants with a capacity of 50 tonnes per day or more unless the nature of the pulp can absorb the work potential. For example, a 30 tonnes per day mill, using a furnish made of 75 per cent straw and 25 per cent softwood will require about 160 HP input for the straw and the same input for the softwood. Ideally, the straw and softwood should be treated separately since disc refiners with such low power inputs are not common. However, for chemi-mechanical pulp production, the work required to finish the semi-cooked pulp is much greater and the disc refiner is therefore the ideal machine. A 400 HP refiner will only produce around 10 tonnes per day, depending on the degree of pre-digestion.

VI.7. Screening

Whatever the nature of the stock or of the end-product, the preparation of stock suitable for use on the paper machine requires screening. As a minimum, stock diluted to the consistency required for the paper machine should be screened just before entering the paper machine flow-box. The function of screening is both to remove undesirable particles and to "clear" the stock (i.e. to obtain an optimum separation of fibres and a uniform suspension of them).

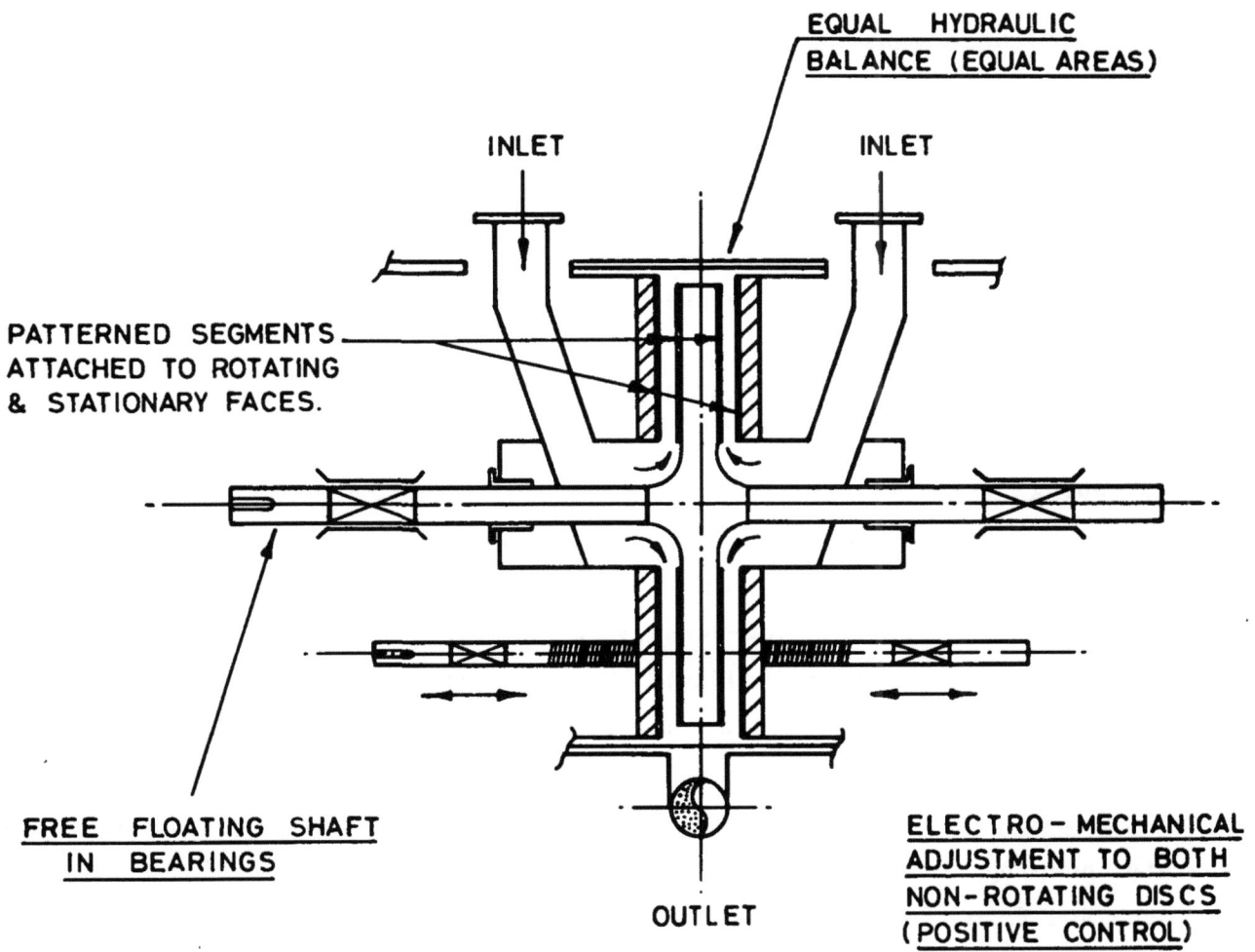

Figure IV 15

DOUBLE-DISC REFINER

Screens can be of varying types: flat vibrating screens, cylindrical vibrating screens or pressurised, fully enclosed screens with rotating foils which induce the flow of material through a stationary screen basket. The latter type of screens has a number of advantages: it is cleaner and does not take in air because it is fully enclosed. Furthermore, it can also be installed directly into the final line to the paper machine, thus simplifying control. For these reasons, pressurised screens have virtually replaced all other forms of screens. They are suitable for capacities of 10 tonnes per day and over. Rejects from these screens can be continuous or intermittent.

Flat vibrating screens are still used, usually as auxiliary screens for rejects from the main screens whenever the quantity is sufficiently large. For very small mills, they may be used as main screens.

Figure IV.16 shows a sectional drawing of a rotary type screen.

VI.8. Cleaning

Cleaners remove sand or grit from the paper stock by centrifugal action (see figure IV.17). It is advisable, for mills manufacturing writing or printing paper to employ low-density cleaners. They operate before the screens, at the consistency applicable to the paper machine (normally, close to 1 per cent). For best results they should have a separate pumping circuit because they are designed to operate at a constant pressure drop which can change if the paper machine grade varies. Many small mills eliminate the separate pumping loop in order to reduce capital outlay. In that case substance and formation control for the paper machine becomes more difficult.

Cleaners can also be of the thick-stock type (i.e. they can function at consistencies of 4 per cent). It is more economical in terms of capital cost and power consumption, but cleaning is less effective. There are also large capacity cleaners operating at around 2 per cent consistency. They are chiefly used for processing waste paper. Cleaners designed to operate on stock of high or medium consistency normally have built-in chambers which receive the collected dirt. Low density cleaners have continuous rejects and may be staged (up to three stages) for fibre economy.

Low density cleaning improves paper quality and reduces wear on parts of the paper machine (e.g. machine wires, foil or vacuum top surfaces).

Figure IV.18 shows a simplified flow diagram for a two-fan pump by-pass system used in conjunction with a three-stage centrifugal cleaning system.

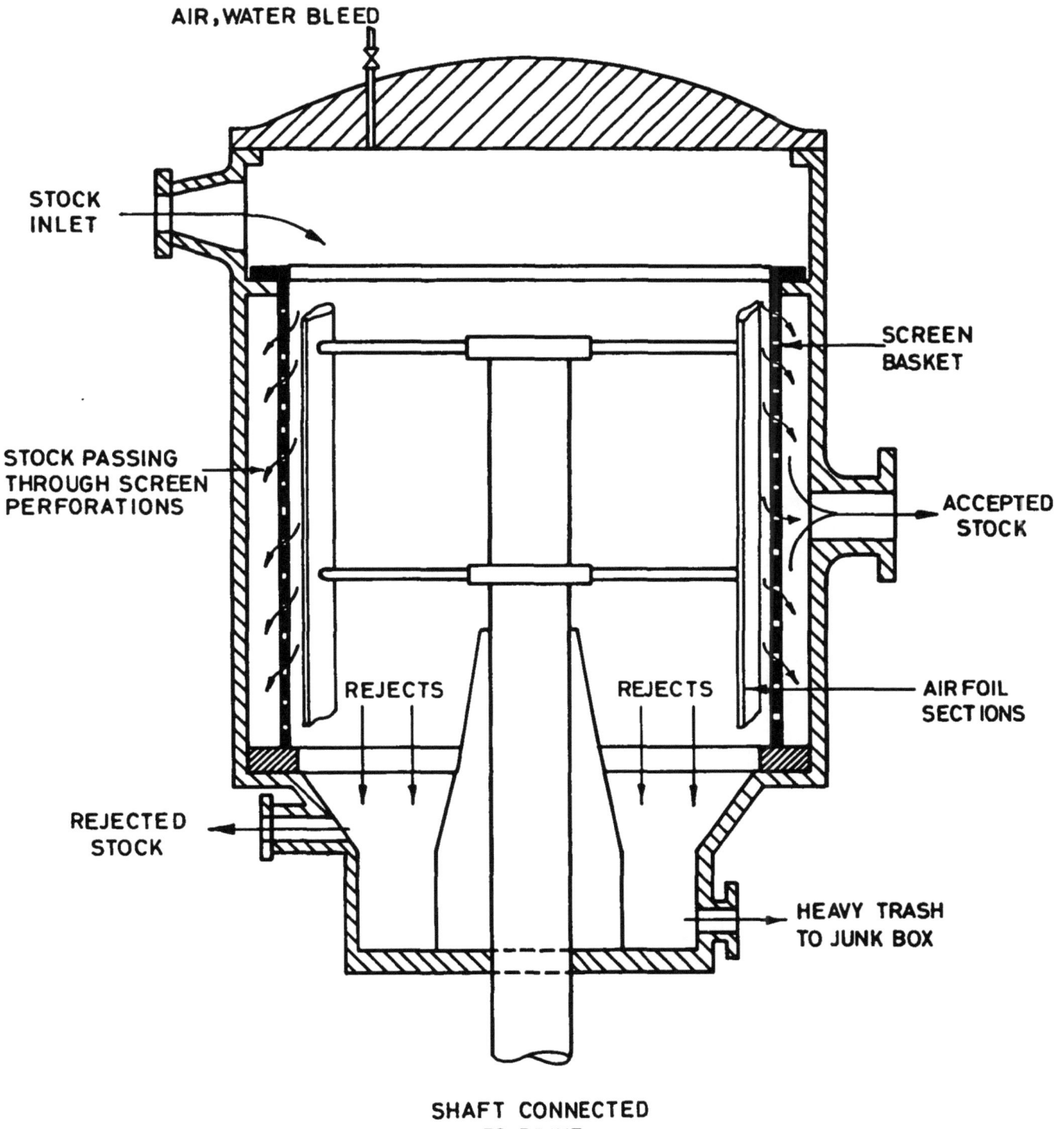

AIR, WATER BLEED

STOCK INLET

STOCK PASSING THROUGH SCREEN PERFORATIONS

SCREEN BASKET

ACCEPTED STOCK

REJECTS

REJECTS

AIR FOIL SECTIONS

REJECTED STOCK

HEAVY TRASH TO JUNK BOX

SHAFT CONNECTED TO DRIVE

Figure IV.16

ROTARY TYPE SCREEN

ILLUSTRATIVE SECTION

LEGEND

1. <u>REJECT LAYER</u> - MOVES IN A DOWNWARD SPIRAL AT CONE WALL.

2. <u>INLET LAYER</u> - MOVES IN A DOWNWARD SPIRAL.

3. <u>ACCEPT LAYER</u> - MOVES IN AN UPWARD SPIRAL.

4. <u>AIR</u> - MOVES INTO AIR CORE.

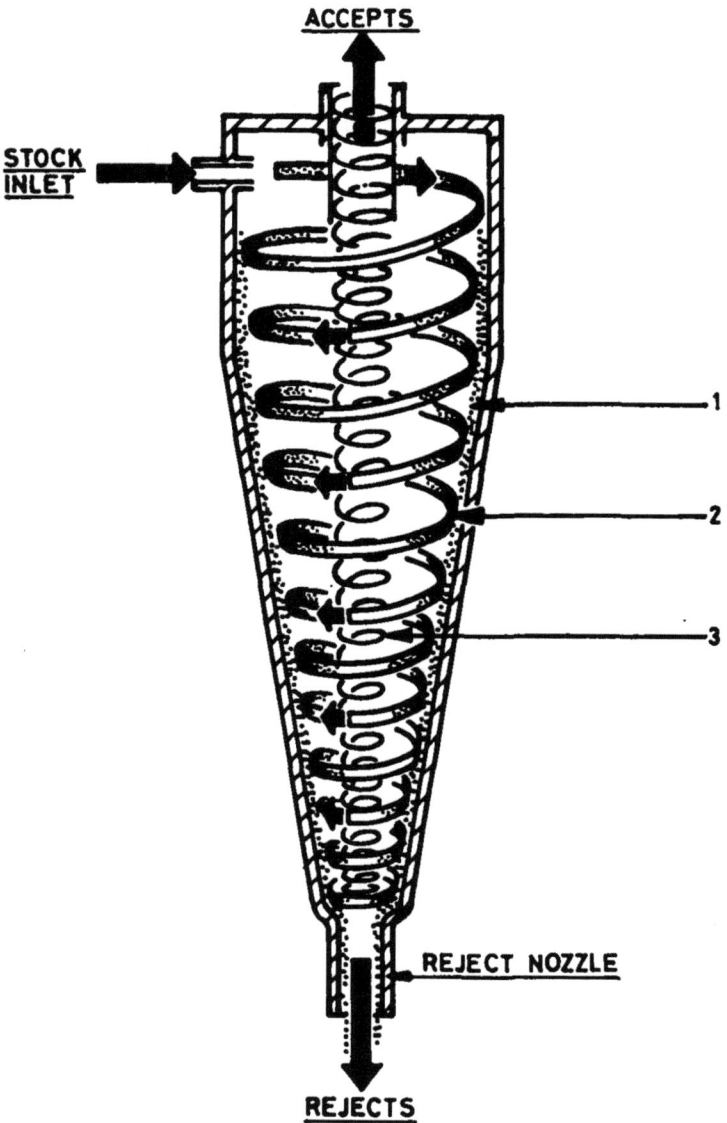

Figure IV.17

CENTRIFUGAL CLEANERS
(GENERAL PRINCIPLES OF OPERATION)

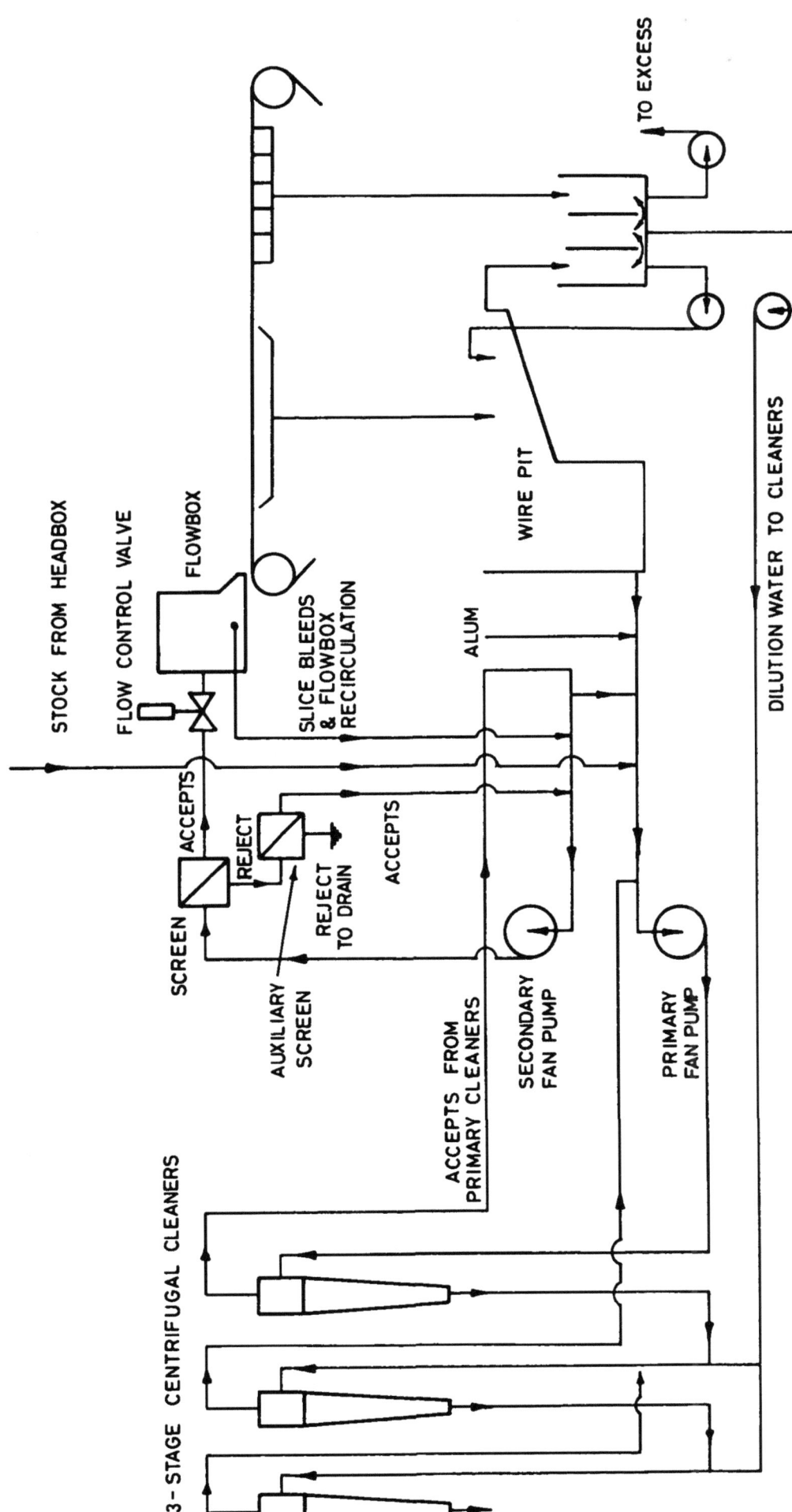

Figure IV.18

SIMPLIFIED FLOW DIAGRAM FOR THE TWO FAN PUMP BY-PASS SYSTEM

VI.9. Waste paper cleaning and screening

If imported pulp, or pulp which has been previously cleaned in the pulp mill, are the furnish constituents, screening and cleaning can safely be limited to the use of the final screens and cleaners described above. On the other hand, if waste paper is used the proportion of dirt and contraries it contains requires more effort and equipment.

For small mills, the pulper is unlikely to be large enough to operate continuously with the trailing rope or junk-remover. For these mills, pulping will generally be carried out on a batch basis. Contraries too large to pass through the perforations remain in the pulper and must be manually removed periodically. The disintegrated paper passes through the perforations to the dump chest, and carries with it such things as sand or dirt, incompletely disintegrated paper particles, bits of string and pins. If the trash content is high, it is best to pump the stock through a thick stock or a medium density cleaner in order to remove particles which can cause undue wear to de-flakers or refiners. If the degree of rejects is tolerable, it is preferable to use the de-flaker directly because it improves the performance of the cleaners which follow. De-flaking should occur as early as possible after pulping. After de-flaking, stock may be subjected to pressure screening in order to remove rejects continuously and to avoid blockage. The rejects can be screened over a simple vibrating flat screen and the remainder of the stock returned to the pulper or to the chest supplying the de-flaker. Refining follows screening and the stock can then pass, via low-density cleaners and a final pressure screen, to the paper machine.

The procedure described in the previous paragraph represents a good practice which is appropriate to small mills. In low wage developing countries, waste is often sorted by hand and obvious contraries or unsuitable material removed. Cleaning may then be restricted to the use of high-density units. Screening may also be limited to the use of the cheaper flat, vibrating units. However, only the lowest grades of paper can be produced with such equipment, and in the long term the savings are likely to be illusory.

VII. STOCK CONTROL

In stock preparation, it is necessary to control two main variables: the degree of refining or beating and the uniformity of stock consistency prior to delivery to the paper machine. The former variable affects paper quality and the latter the paper substance variations, which, in turn, can affect the efficiency of operation. The degree of refining can be assessed by motor loads for refiners or by human judgement for beaters. Large-scale mills may have automatic refiner control, while small mills may rely on the judgement of the paper machine operator. Variations in freeness are not usually spasmodic or short-cycled: there is a time lag which is dependent on the storage chest capacity provided that the refining or beating load is constant. Consistency variations are almost impossible to avoid without control and consistency regulators are indispensable for large mills. Theoretically, consistency control is inherent in batch processing. In practice, it depends on the human factor and also on a realistic measurement of moisture content in the in-going material. This measurement is, however, very difficult to obtain.

Consistency regulators

There are many types of consistency regulators. The modern ones emit an electrical or pneumatic signal to a recording chart, or in sophisticated versions to a computer. They are seldom in use in small mills in developing countries because of their high price. Furthermore, the operators seldom understand their functioning or can maintain and calibrate them. The operating efficiencies in many such small mills are much lower than necessary, chiefly on this account. In the majority of cases, the probable reason is that only one unit has been installed and consistency variations are out of range most of the time. A consistency regulator can only add water in order to adjust the stock thickness. Control is better with two regulators in operation. Large, sophisticated mills have several consistency regulators, one at each practical point of application. Thus, the final unit has only a very small variance to control.

Small mills in developing countries are unlikely to have a trained instrument supervisor required by modern regulators. There are, however, older, more mechanical units still available (e.g. the Salle unit). It is recommended that the mill manager should install at least two suitable units and persevere until reliable operation is obtained. The rewards in paper machine efficiency and uniformity of quality through better refining are substantial and will certainly justify the additional investment.

Figure IV.19 shows a simple stock control system with recommended consistency regulation.

- 85 -

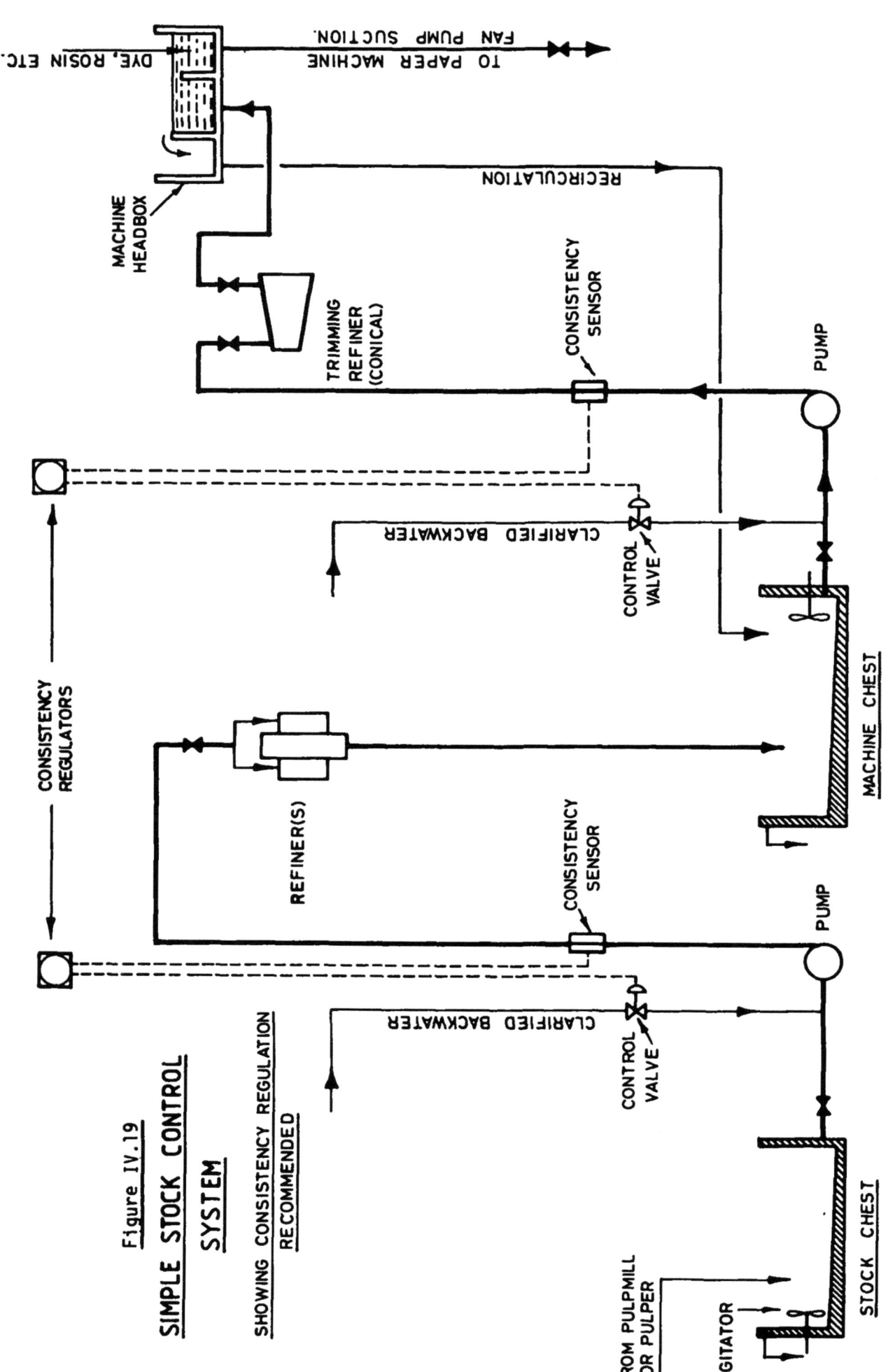

Figure IV.19

SIMPLE STOCK CONTROL SYSTEM

SHOWING CONSISTENCY REGULATION RECOMMENDED

CHAPTER V

THE PAPER MACHINE

The transformation of stock into paper or board is carried out through four main operations:

- Formation and drainage;
- Pressing;
- Drying; and
- Reeling, winding and sheeting.

This chapter provides a brief description of these four operations, including that of the equipment associated with each one and of the various chemicals which are added to the stock. The description emphasises processes and equipment of interest to small mills in developing countries. Variations in these processes for the production of board are also provided.

I. PAPER PRODUCTION

I.1. Formation and drainage

This processing stage consists in distributing the fibre and water suspension evenly over a wiremesh so that when the water drains through the mesh a wet sheet of paper is retained. In the case of hand-made paper, the wiremesh is horizontally stationary while being manually agitated in order to ensure evenness of the formed sheet of paper. In the paper machine, the wiremesh carries the suspension along horizontally at various speeds, depending on manufacturing conditions.

Hand-made paper is produced in the form of individual sheets, the size of the sheets being limited by the ability of the operator to lift and agitate a wire-bottomed frame after dipping it into a vat containing a suspension of prepared fibres in water.

Machine-made papers are formed continuously on a moving wire, or between two synchronised moving wires, and are produced in wound reel form. In the standard Fourdrinier type machines (see figure V.1), the stock is allowed to flow on to the wire. A flowbox is used to ensure that it is evenly spread,

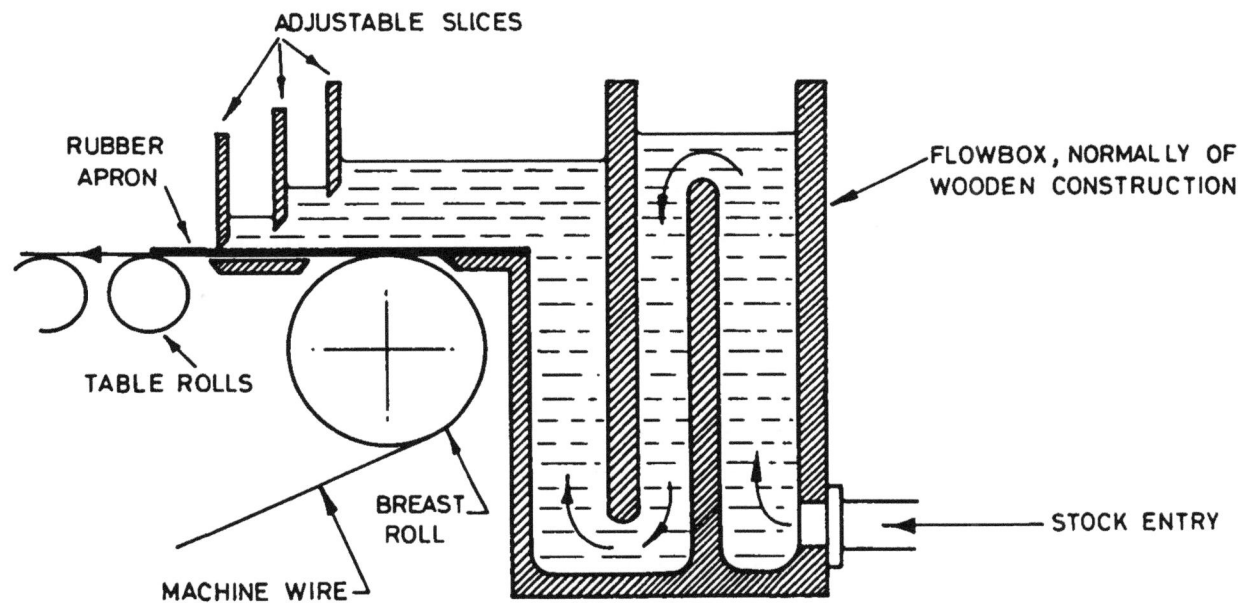

ADJUSTABLE SLICES

RUBBER APRON

FLOWBOX, NORMALLY OF WOODEN CONSTRUCTION

TABLE ROLLS

STOCK ENTRY

BREAST ROLL

MACHINE WIRE

SLOW-SPEED FLOWBOX & APRON SLICE

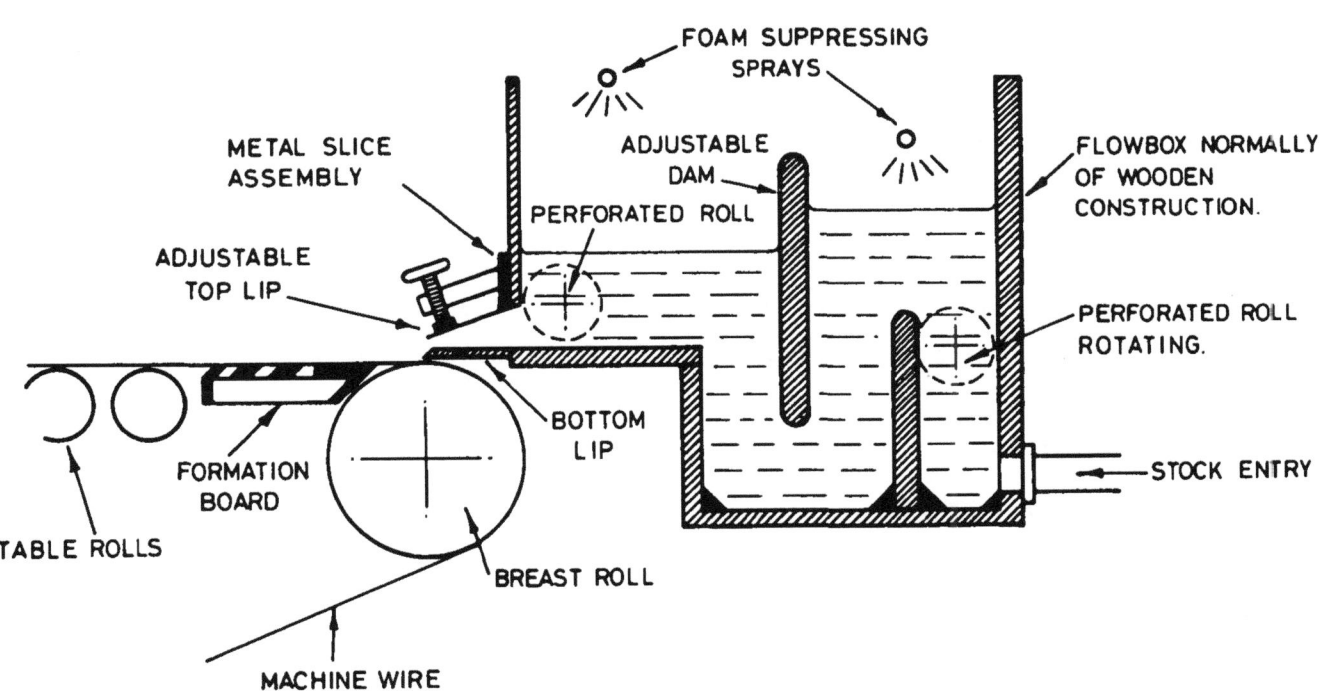

FOAM SUPPRESSING SPRAYS

METAL SLICE ASSEMBLY

ADJUSTABLE DAM

FLOWBOX NORMALLY OF WOODEN CONSTRUCTION.

PERFORATED ROLL

ADJUSTABLE TOP LIP

PERFORATED ROLL ROTATING.

FORMATION BOARD

BOTTOM LIP

STOCK ENTRY

TABLE ROLLS

BREAST ROLL

MACHINE WIRE

FLOWBOX FOR SPEEDS UP TO 200 M/MIN
PROJECTION SLICE — GRAVITY TYPE.

Figure V.1

FLOWBOXES FOR SMALL-SCALE MACHINES.

the fibres kept in suspension, and that the hydraulic speed of the fibre and water slurry leaving the flowbox is equated with that of the travelling wire. As the machine speed increases, the hydraulic head in the flowbox must be increased in order to give an equivalent jet velocity.[1] For example, a speed of 300 m/mn requires a height of 130 cm. At very low speeds, an "apron" slice is used (a rubber apron resting on the wire while the stock is poured over it). The flow is controlled by an adjustable dam called a "scissors" slice located behind it. For speeds below 180 m/mn, the flowbox has an open top, and the stock flows by gravity onto the wire through a projection slice opening. The top lip can be deformed slightly at intervals across its width in order to control the flow and ensure evenness of deposition. The gravity flowbox (see figure V.2) is inexpensive and can be locally made, except for the slice and perforated rotating rolls which are fitted to maintain dispersion. It is suitable for speeds around 200 metres per minute and therefore appropriate for small mills.

Gravity flowboxes are impractical for speeds much above 200 metres per minute in view of the height required in order to provide the jet velocity. Consequently, pressurised flowboxes are used instead. Pressure is exerted pneumatically, hydraulically or by a combination of both. Pressurised flowboxes are very expensive and need costly and sophisticated control. They are therefore not appropriate for small mills in developing countries. The latter should not, therefore, use paper machines with an output exceeding 200 metres per minute.

For very low speeds, boxes with a fan-induced vacuum compartment are used in order to increase the stock depth. These boxes are not expensive and are simple to operate. They can assist in formation control by increasing the stock height without correspondingly increasing jet velocity.

Formation for hand-made papers is "square" (i.e. the fibres are not oriented to deposit in any given direction). On the other hand, machine-made papers are oriented in the direction of the paper machine. Consequently, the paper properties lengthwise are not quite the same as those crosswise. Some control over this difference in properties can be exerted by adjusting the velocity of the stock through an increase or decrease of the pressure head, thus providing what is termed a "lead" or a "drag".

---

[1] The relationship between the jet velocity and the hydraulic head follows Newton's law of gravity: $V^2 = 2 gH$

where: V = speed in m/sec.

g = gravitational acceleration in m/sec/sec.

H = height of the liquid in the flowbox, in m.

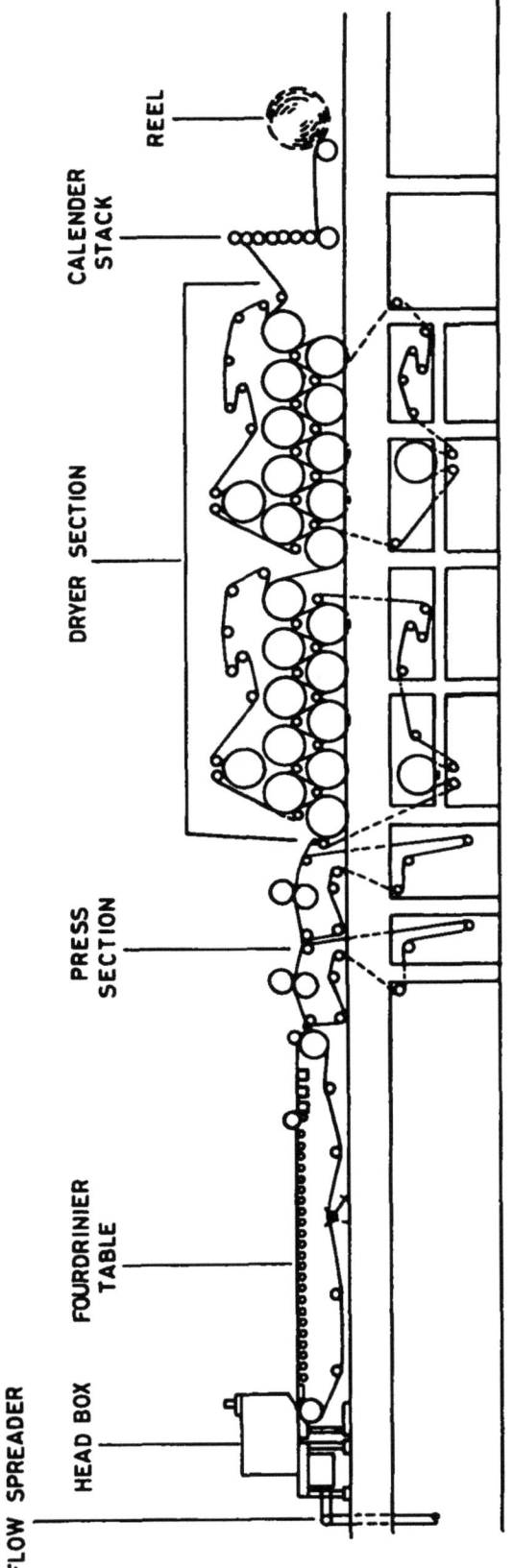

Figure V.2

SIMPLE FOURDRINIER MACHINE WITH GRAVITY FLOWBOX.

The stock goes onto the wire at a consistency which depends on the type of output and the furnish used. Generally, thin papers need low consistencies (down to 0.25 per cent for tissues) while for heavier papers, higher consistencies (up to 1.25 per cent) can be used. Control is exerted by regulating the water added to the metered stock, before the screens.

The separation of the formed paper from the wire requires that it supports itself across a gap. Thus, the water content must be reduced in order to give the paper the required wet strength. The consistency must be increased to a minimum of 12 per cent, or preferably higher. The wire is supported by small rolls, or foil surfaces which rupture the water film and induce a vacuum that promotes drainage. The vacuum effect is a function of speed, reaching barometric limits at high speeds in machines equipped with rolls. The foil was introduced in order to limit this effect by inducing half the vacuum intensity of a rotating roll since a high vacuum intensity can impair formation by unduly disturbing the forming paper. In order to keep the wire length reasonably short, air suction is created through perforated or slotted boxes fixed underneath the wire and sited in such a way as to consolidate the paper when it is sufficiently set. In modern plants, a rotating, perforated roll with an internal suction box located after the suction boxes is added. This additional feature further removes water and improves wet strength.

Suction rolls are expensive and use a great deal of energy. Small mills are, however, advised to use them because efficiency is substantially improved. Plain, felt-jacketted rolls, running above another roll beneath the wire, were employed before the introduction of the suction roll and are still used for very small capacity machines in order to reduce capital and operating costs. However, investment in suction rolls is justified for machines with a capacity of 20 tonnes per day and above.

When the wet-formed sheet is separated from the wire, a tension is imposed on it and there is a risk that the paper may break. The jacketted couch roll induces a slight vacuum which lifts the sheet at low speeds. A strip can be hand-plucked from the felt surface and thrown onto an adjacent felt which takes it to the presses. The strip can be widened to a full width after it has gone through the machine. The tension required to peel the paper from the wire increases with speed while the strength of paper required to take the strain is lower for thin paper than for heavier grades. "Pick-up" devices have therefore been developed. For tissue grades of paper, a "lick-up" may be used: it involves the press felt running in synchronism with, and touching the forming wire above an "open" roll in order to induce a vacuum which causes the paper to adhere to the felt. Papers which can be lifted, full-width, in this manner, must have a substance which cannot exceed 45 grams per square metre (45 g/m^2). Consequently, other methods are necessary for high speeds and heavier grades of paper. The "vacuum pick-up" was developed for this purpose. It operates as follows. A suction roll, running inside a felt which also runs through the following press, is adjusted in such a way as to touch the forming wire. The wet sheet adheres to the felt and leaves the wire full-width. Vacuum pick-up arrangements are very expensive and involve sophisticated speed synchronism, adjusting controls, etc. A sophisticated pick-up is not necessary for small capacity mills. A less expensive pick-up can be used effectively for speeds below 300 metres per minute. It removes the sheet at, or near, right angles to the wire, that being the angle at which the tension requirements for removal are minimum. A plain, forward wire roll must be added to the wire arrangement in order to provide the angle required. Sometimes a small, slightly pressurised box (called a Baggely box), is added in order to facilitate the removal of the sheet. However, it is not essential. Careful positioning of the roll (which carries the press felt) in relation to the wire should be sufficient.

I.2. Pressing

The formed paper leaving the wire is 12 to 18 per cent dry, depending on the machine characteristics and the nature of the furnish. Its water content is therefore six or more times higher than that of the fibre. Most of this water must be removed before the paper can be reeled. Less energy is required and the paper quality is improved if a maximum of water is expelled mechanically, by pressing. A press comprises two rolls revolving in contact, with pressure applied between them. There should be at least two sequential presses in a paper machine. The first press is normally double-felted (i.e. each roll runs inside an endless felt so that the paper is squeezed between the two felts which absorb the expelled water, leaving the sheet drier and stronger). The pressure must be limited to that which the wet paper can sustain. If too great a pressure is applied, the paper can crush and be spoilt. This is why successive presses are used in order to reach an optimum pressure by stages. The second press normally has a single felt around the bottom roll, which is rubber covered and softer than the top roll. The latter is hard, and is ideally of granite, though more frequently made of a hard but porous composition such as Stonite. The pressed paper adheres to the top roll from which it is removed as a strip by air jet. The strip is widened to full width after it has reached the end of the paper machine.

A third, smoothing or reversing press may be incorporated for the production of writing or printing paper. The smoothing press has no felt, and both rolls are hard. Bronze is often used for the top roll. The function of the press is not so much to remove water as to improve flatness and the paper finish. The reversing press has a bottom felt, but the paper is turned and fed into the opposite side of the nip after which it is turned again to move forward in the proper way. The object of this procedure is to have the hard roll applied to the underside of the paper as well in order to equalise smoothness on both sides of the paper.

Figure V.3 shows the pressing section of a typical paper machine used for the production of 5 tonnes per day of tissue and MG Kraft papers.

Presses

Presses may be plain (i.e. both top and bottom rolls being undrilled) or grooved. They may also be suction presses on which the bottom roll is perforated and has an internal vacuum chamber, or may be grooved. Plain presses are particularly appropriate for small mills because they are fairly cheap and require little skilled maintenance. At low speeds (100 metres per minute or less), the plain press performance is almost as good as that of the more expensive types in view of the long pressing time in the nip area. The performance of plain presses can equal that of the best if the fabric press principle is adopted whereby a relatively strong, open-meshed, endless plastic belt runs between the press felt and the roll. An even simpler device makes use of special high-duty ribbed felt. The performance of a plain press equipped with this felt can exceed that of the large, fast machine in view of the narrower width of small paper machines.

It is strongly recommended that granite top rolls be used in view of their greater efficiency when manned by inexperienced operators. It is necessary for most small mills in developing countries to import the special felt fabrics and the granite rolls. The loading of presses can be carried out with pneumatic or hydraulic devices, or by hand. The latter method is suitable for the small, slow paper machine and does not require any capital investment.

The moisture content of the paper leaving the press section can be as low as 50 to 60 per cent (depending on the paper grade), if high-duty presses are used. The latter can withstand nip pressures of about 600 lb/linear in. of width. While that pressure may be easily achieved in small paper machines

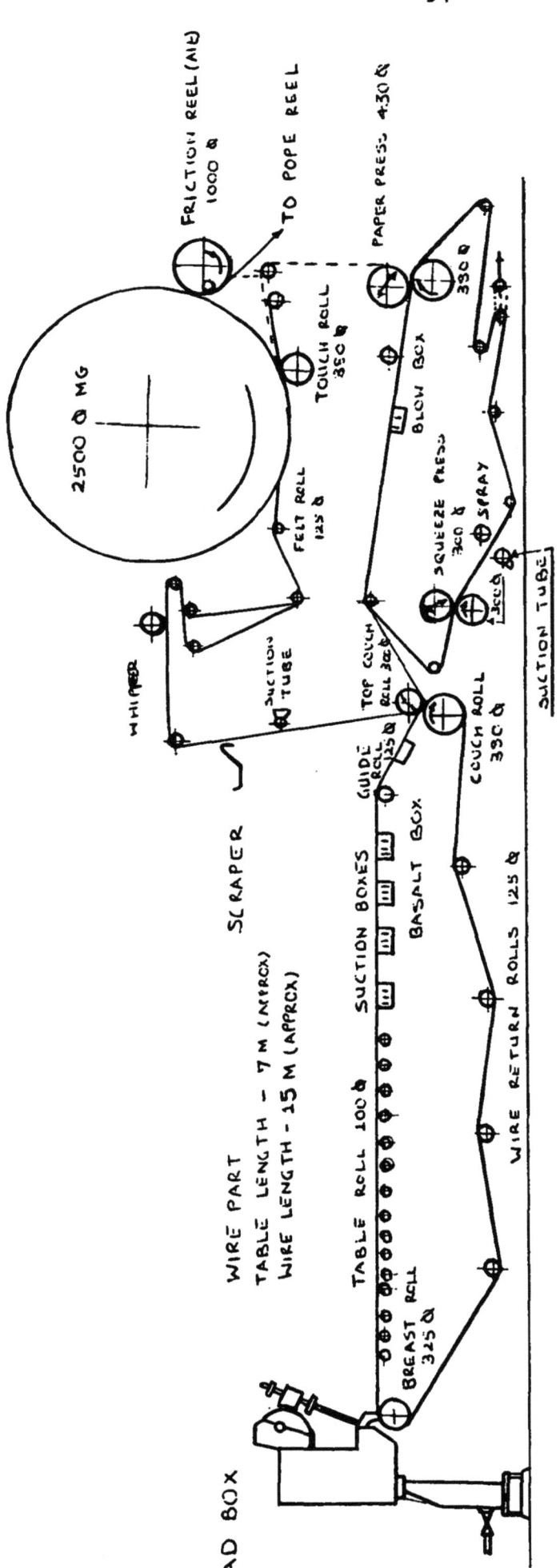

Figure V.3

Typical flow scheme for 5 T.P.D. capacity paper machine for tissue and MG Kraft

machines, it requires special materials and precision-cambered grinding and control in the case of large machines. It is now considered necessary for wide, fast machines to have crown-compensating press rolls (i.e. rolls which can be made to expand in order to provide a cambered surface which, under pressure, becomes a horizontal nip for equal loading over the sheet width). The small machine does not need this expensive sophistication.

Good pressing does more than reduce the amount of steam required for drying. The paper is strengthened by good, even pressing and fewer breaks occur. The surface is also less fibrous, so that the dryers stay clean longer and operate more efficiently. Experience shows that the major cause of the inefficiency of some small mills is the lack of good pressing. This is paradoxical in view of the fact that small paper machines should obtain better results from pressing than can be achieved by large, fast machines.

I.3. Drying

After pressing, the paper still contains 50 to 60 per cent of moisture, most of which must be removed before reeling. Application of heat is required for this purpose. The most common drying method makes use of rotating cast-iron cylinders that are steam-heated internally. Plain or machine-finished paper is normally dried on a series of such cylinders, depending on the quantity of paper being processed. The standard diameter of cylinders is approximately 1.5 m; smaller or larger cylinders can be obtained, but are seldom used. Cylinders must be capable of sustaining one of two standard internal steam pressures: up to 5.27 kg/cm^2 (75 p.s.i.)[1] or up to 8.79 kg/cm^2 (125 p.s.i.). Other pressures are less common. The rate of drying increases proportionately with higher pressures and temperatures. The external surface of cylinders has a ground finish while the internal surface is machined in order to obtain an even thickness, and therefore equal drying, across the width.

Each dryer is fitted at the back with a nozzle which combines the functions of steam admission and condensate removal. Large, wide cylinders may have such a nozzle at each end in order to ensure a more even drying width-wise. A more uniform drying is achieved by narrow machines. The nozzles connect to a steam distributor pipe located inside the cylinder, and to a collecting syphon for the removal of the condensate. Syphons may be stationary (i.e. they remain in one position near the bottom of the inner surface) or of the revolving type. In this latter case, they rotate with the cylinder and remain in one position relative to the inner surface. Usually,

---

[1] Pounds per square inch.

stationary syphons have an internal bearing which, over a period of time, can cause problems. They are not recommended for the small paper machine, although their performance is better at speeds below 300 metres per minute. Above this speed, the condensate "rings" the cylinder (i.e. it is held against the inner surface and revolves with it). At lower speeds, it remains in a pool near the bottom, so that a revolving nozzle collects it at each revolution while the stationary nozzle removes it continuously. However, internal bearing wear in stationary syphons can adversely affect the gap between the cylinder surface and the syphon, and ultimately damage the internal fittings.

It is important to remove the condensate at a rate equal to that of formation because the power required to drive the cylinder can be several times greater if it is waterlogged. Large-capacity machines, with a multiplicity of dryers, have special, three-stage systems for steam supply and condensate removal. They use flash steam from full pressure dryers in order to supply a number of dryers at lower pressure. Flash steam from the latter is then exhausted under vacuum in order to supply the first dryers which should be kept at a lower temperature in order that the paper will not stick. Automatic pressure regulation between stages is incorporated, and overall drying control can be added. Such drying systems are expensive and need expert supervision. They are not required for small machines with a limited number of dryers. A steam trap on each cylinder equipped with an air release, is more commonly used.

The dryers are driven through a train of gears mounted on the back. Fast running, large-capacity machines need fully enclosed gears with a constant circulation of lubricating oil. Supporting bearings for these machines must also be of an anti-friction type, and force lubricated. For the small, slower machine, open, grease-lubricated, much less expensive gears are sufficient. Furthermore, anti-friction bearings are not essential (an advantage in developing countries), and lubrication can be done by individual oil-drip, or, for very slow machines, by solid grease blocks.

The paper is held to the dryer surface by endless upper and lower felts, which are driven by the cylinders. These felts were originally made of cotton-asbestos weave. For the large, modern machines, they have been virtually replaced by "screens" which are made of strong, fairly open plastic mesh. The mesh is more expensive but lasts longer, and gives an improved drying performance for the following reasons: the screens allow steam from the paper to escape more readily, do not need separate drying and carry air into the spaces between the cylinders, thus facilitating the removal of moisture-laden vapour. For the small machine, with a narrower width, dryer screens are not essential, but may still have advantages which justify the additional expense.

The feeding of paper through the dryers of a large, fast machine equipped with a large number of cylinders would be impossible without Sheahan ropes. These are a pair of endless ropes running together through the machine, following the paper run. They are carried in a special ring bolted to the front of each cylinder, and return over a series of pulleys under the machine. The tail of the sheet, from the presses, is laid between the two ropes which hold and convey it to the last cylinder. For speeds of 200 metres per minute or below and a relatively small number of cylinders, Sheahan ropes are not essential. For these machines, the tail of the sheet can be hand-fed.

For the large capacity machine (see figure V.4), a fully enclosed hood over the dryers (extending right down to the basement) is essential for two reasons: it provides a balanced pressure in the dryer zone for uniform drying, and it protects machine operators against excessive heat and humidity. High-duty exhaust fans and humidity control are incorporated. Automatic paper-break detection equipment is essential. The small machine needs, at best, a simple semi-open hood and exhaust fans. Thus, for high capacity machines, the dryer section must embody expensive refinements which are not necessary for the smaller, slower units. This is one respect in which unit capital costs are significantly lower for small machines.

The foundry requirements for cast-iron cylinders are of a special nature, and the equipment for machining and grinding them is fairly expensive, and may not be available in most developing countries. For these reasons, the local manufacture of cylinders may not be economically feasible. In most cases, they must be imported. Rolled, welded steel cylinders are sometimes made in developing countries, whenever imports are restricted. No internal machining is required, and they can sustain the internal pressures more simply. Rusting may take place when the machine is stopped for a long period, unless adequate precautions are taken. In general, however, good performance may be expected from cylinders such as these.

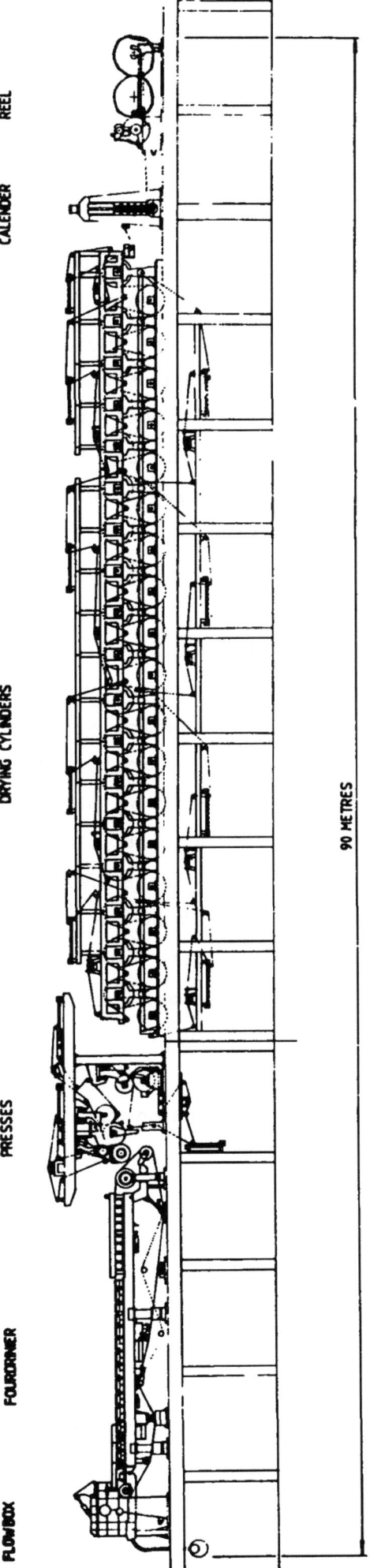

FLOWBOX FOURDRINIER PRESSES DRYING CYLINDERS CALENDER REEL

90 METRES

Figure V.4

HIGH – SPEED, LARGE CAPACITY FOURDRINIER MACHINE WITH PRESSURISED FLOWBOX AND VACUUM PICK – UP SECTION.

Any paper shrinks, lengthwise and crosswise, as it dries. The shrinkage is greater for well-beaten or refined grades than for other grades. To accommodate shrinkage as well as to limit the accumulation of broke during breaks or feeding-up, dryers in large machines are sectionalised, thus limiting the number of dryers in any section to a practical number (from eight to 16). Each section has an upper and lower felt, and each felt needs a guide to keep it on the cylinder surface, in addition to a stretch gear in order to accommodate stretch. The gear normally incorporates an automatic tension device. A machine with a capacity of 30 tonnes per day can incorporate all the dryers required in one section, whereas a large machine requires at least three or more sections. The cylindrical dryer flattens the sheet (a desirable function) and is economical in the utilisation of heat. Other forms of drying are sometimes used as auxiliaries, but are unlikely to be of practical use in developing countries.

I.4. Reeling, winding and sheeting

It is not desirable for paper to be completely dried. A moisture level of 5 to 8 per cent should be retained for the paper to be in equilibrium with the atmospheric moisture. A paper with this standard degree of moisture is said to be "air-dry". After drying, the paper is reeled on the machine for subsequent finishing. Writing and printing grades of paper must first pass through the rolls of a calender in order to acquire a smooth surface. To achieve very high standards, two successive calenders may be used. On some machines, the last cylinder in the dryer train may be water-cooled rather than heated: it is called a "sweat-cylinder" because, in the humid atmosphere surrounding it, condensation forms on its surface and is imparted to the paper before the calender, thus improving the paper finish.

A calender and pope reel arrangement are shown on figure V.5.

Size press

For top quality writing and printing papers, a size press may be installed after the main dryers. It consists of a two-roll unit which may be vertical, horizontal or inclined. The paper passes through a solution of starch and water between the rolls which squeeze out the surplus liquid. Since the paper is re-wetted by this process, it must be dried again. Consequently, after-dryers, corresponding to approximately one-third of the first dryers, are added before calendering.

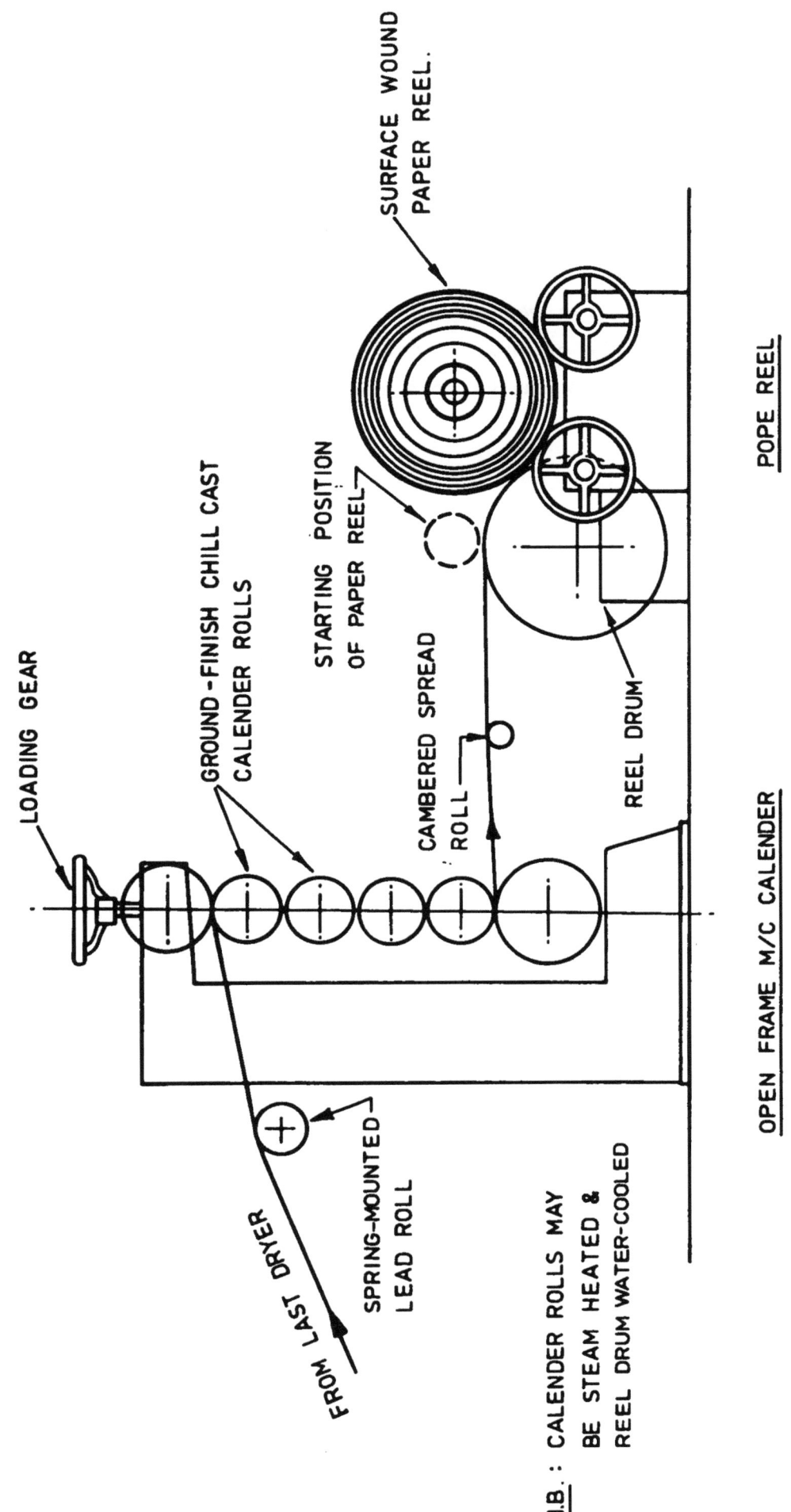

LOADING GEAR

GROUND-FINISH CHILL CAST CALENDER ROLLS

STARTING POSITION OF PAPER REEL

SURFACE WOUND PAPER REEL.

CAMBERED SPREAD ROLL

REEL DRUM

FROM LAST DRYER

SPRING-MOUNTED LEAD ROLL

N.B. : CALENDER ROLLS MAY BE STEAM HEATED & REEL DRUM WATER-COOLED

OPEN FRAME M/C CALENDER

POPE REEL

Figure V.5

CALENDER & POPE REEL ARRANGEMENT.

For large capacity machines, it is necessary to install a broke pulper beneath the calenders and last dryer because the amount of broke involved during feeding-up or after a break is too great for physical handling. Small machines do not need such a unit: broke can be manually handled and taken back for repulping.

Reels

At the end of the machine, the paper is reeled up. Large machines use sophisticated reels, pneumatically or hydraulically loaded. For small paper machines, a simpler and much less expensive pope reel may be used (see figure V.5).

Winders

After reeling, the reel of paper is lifted from the paper machine and taken to a winder. The purpose of the latter is to slit the paper into commercial widths and to repair breaks, if any, in order to obtain a continuous reel of paper. The latter is then wrapped and despatched, or processed further. The winder operates intermittently, producing sets of smaller diameter, uniform sized reels from the full-width, large diameter machine reel. It must therefore operate at a higher speed than that of the paper machine. A speed two to three times higher is usually required. For the large, fast machines, speeds above 2,000 metres per minute are necessary. These high speeds require special design, drive and operation. The size and weight of the parent reels and that of the final reels are such as to require special lifting apparatus, built-in reel ejectors and mechanised reel bar removal. Thus, the large, fast winder is also a sophisticated and expensive machine. For the small mill, the winder is slower and much less expensive since it can be operated manually.

It is normal for writing and printing papers to be cut into sheets before despatch to the customer in the form of reams or in smaller "cut-sizes". For developed countries, the proportion of sheets to reels normally exceeds 60 per cent. For developing countries, the proportion should be higher (around 80 per cent) because reel-fed printing machines are less common. Wrapping grades should also include a large proportion of paper in sheet form.

Cutters

A cutter is used for the cutting of paper into sheets. For large capacity mills, the cutter has become a highly sophisticated, expensive machine equipped with such features as built-in inspection and rejection devices and automatic counting. For the small "cut-size" papers, special cutters with built-in packaging are used. The small mill can make use of a more simple cutter. In developing countries, manual sorting, packing and wrapping is feasible in view of the low wage level and is preferable because it creates employment. The small, "cut-size" sheets may be produced with a simple, manually operated guillotine.

II. THE BOARD MACHINE

II.1. Formation

In developing countries, the wet end of board machines is almost invariably of the simple multi-vat type. Other boardmaking units are available, but usually have a high capacity, and are specifically designed for operation at speeds beyond the range of the vat. In its simplest form, the through-speed in the vat is limited to around 100 m/min. There are two types of vats: the counterflow vat and the uniflow vat (see figure V.6). In the former type, the stock enters the vat and moves in an opposite direction to that of the cylinder mould rotating within it, whereas in the uniflow vat, the stock flows in the same direction as that of the mould. Counterflow vats tend to add a greater quantity of stock per vat but the formation from a uniflow vat is superior in quality. Thus, it is common for both types to be incorporated in the same machine. Uniflow vats are used for the surface layers in order to improve appearance and printing quality while counterflow vats are used for the middle layers. As many as nine vats can be used, depending on the thickness of board to be produced. For a given thickness, the quality improves with the number of vats employed. The small mill is unlikely to use more than six vats in view of the high capital costs and because extremely high quality boards are not widely marketed. Fair quality board may be produced with a board machine equipped with six vats.

For most products, two grades of stock are used: bleached chemical pulp for the liner and mixed waste for the remainder, the latter representing the bulk of the board. Thus, a board mill is predominantly a waste-based mill. Operation of the mill is relatively simple. A long, endless felt runs in contact with each cylinder mould and is held in position by a soft rubber couch roll. The felt picks up the formed layers successively, one after another, until the desired number of plies has been added. The board remains

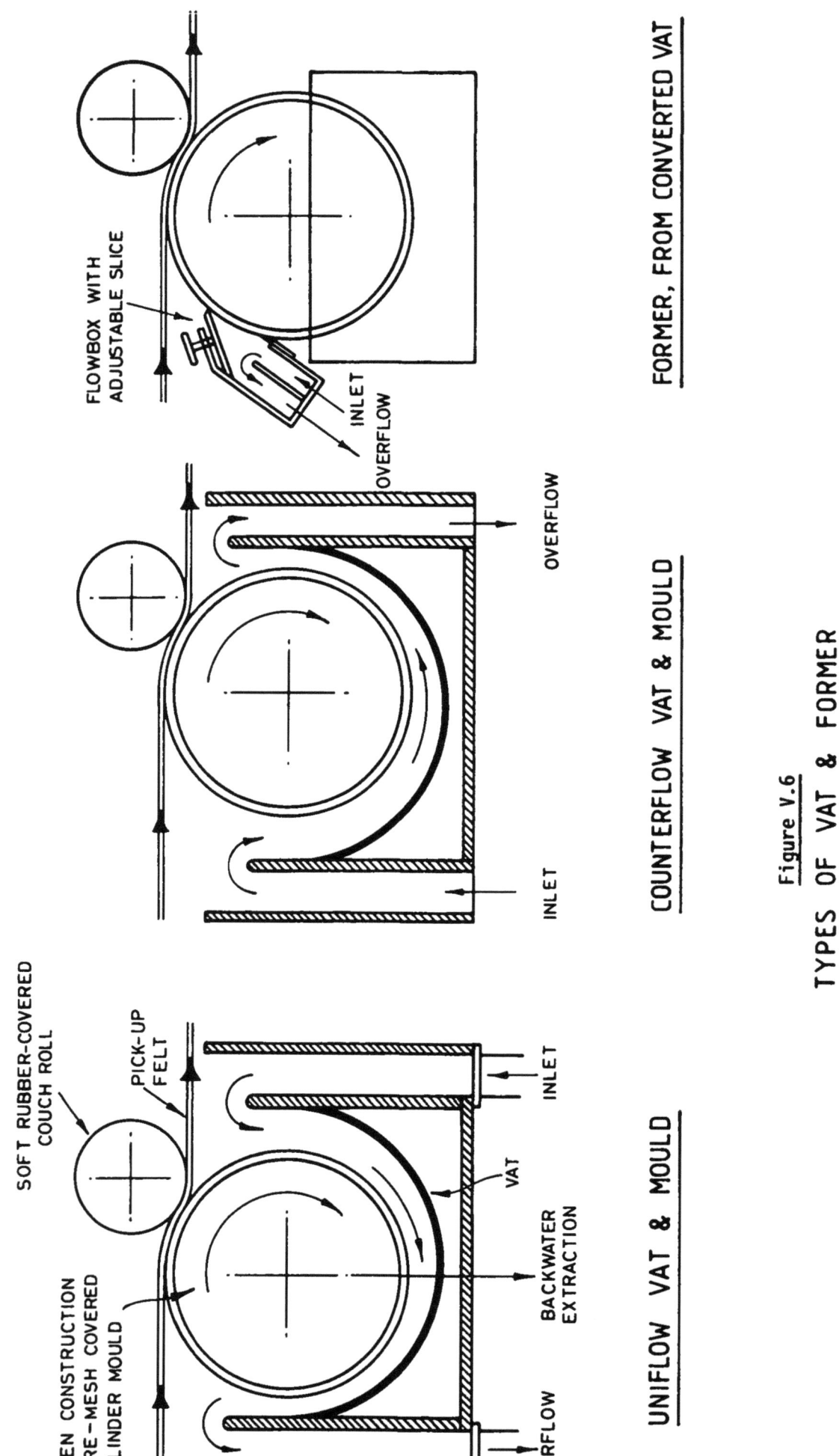

FLOWBOX WITH ADJUSTABLE SLICE

INLET

OVERFLOW

FORMER, FROM CONVERTED VAT

OVERFLOW

INLET

COUNTERFLOW VAT & MOULD

SOFT RUBBER-COVERED COUCH ROLL

PICK-UP FELT

OPEN CONSTRUCTION WIRE-MESH COVERED CYLINDER MOULD

VAT

INLET

BACKWATER EXTRACTION

OVERFLOW

UNIFLOW VAT & MOULD

Figure V.6

TYPES OF VAT & FORMER

on the felt until it has passed through the first press section because it is heavy and wet and unable to support itself until the moisture content has been reduced. The felt runs in opposite direction to the machine until the last vat has been passed. It then turns and runs overhead, in the machine direction, to the presses. The turning roll can be a suction roll. However, for a small capacity mill, it is unlikely to be of the suction type in view of the high capital and operating costs.

There is no control over formation or width-wise substance in the simple vat. For this reason, the "dry" vat, or "former" has recently been introduced. In operation, the cylinder mould does not run partially immersed in stock (hence the term "dry"). Instead, each vat has an individual flowbox with a slice which can be adjusted for level width-wise or for formation in machine direction by regulating the stock velocity. The scope of the vat is thus improved: higher speeds are possible (up to 200 m/mn) and more stock can be added per vat without loss of quality. High paper qualities can also be obtained. These vats offer excellent prospects for small mills in developing countries since with two or three such vats a very wide range of good quality paper and board can be produced from the same machine. The machine is simple to operate, conveys the sheet full width to the presses, is low in power consumption and maintenance, does not use wire guides or stretches, and runs efficiently. It should be added that the normal "wet" vat can also produce, at low speed and narrow width, high paper grade qualities (e.g. banknote paper, filter papers and cotton-based art grades). The inherent formation characteristics suit such papers.

II.2. Board pressing (see figure V.7)

Because board is normally heavy in substance (up to 450 g/m^2 or more), it carries more moisture than paper per unit length. Furthermore, the water is more difficult to remove: pressure must be applied gradually in order to avoid crushing. For this reason, more presses may be needed than in the case of paper. The production of a board machine is usually limited by its drying capacity, which is strongly influenced by the press section. Another very important quality consideration is the ply-adhesion which is also influenced by pressing. "Baby-presses", which are small, plain presses with light loading, are often incorporated with the long felt in the vat section in order to bring the board to a condition suitable for the main presses. Plain, grooved or suction presses can be used. However, nip pressures are normally lower than for paper, and the moisture content is seldom higher than 65 per cent. The small mill may, at first, use plain presses in view of the lower

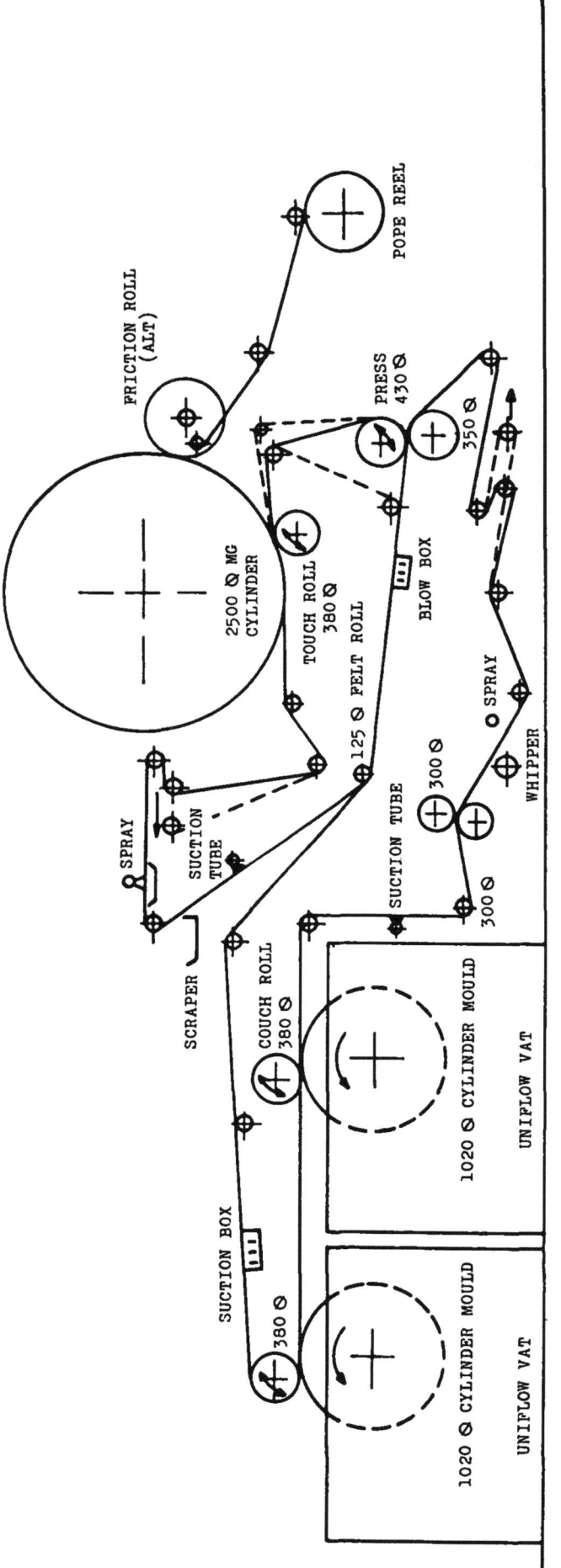

Design and operation data

Type of machine	: Cylinder mould (M.G.)
Capacity (gross)	: 4.5 to 5 tonnes/day
Substance range	: 40 to 120 G.S.M.
Operating speed	: 12 to 70 m/min
Trimmed width	: 1500 at reel
Mould width	: 1750 (effective)

All sizes are in mm.

Figure V.7

5 TPD capacity paper machine with two cylinder moulds
and M.G. cylinder

capital and operating costs. Suction presses, in first and second press positions, can be added later in order to improve quality and increase capacity. Figure V.7 shows the flow diagram of a board machine with a capacity of 5 tonnes per day. The press is shown on the right in the figure.

II.3. Board drying

The drying section of a board machine is similar to that of a paper machine but control may be different. Board is predominantly sheeted, and the sheets should stay flat in order to avoid problems with carton making. For this reason, board machines may have separate steam systems for the top and bottom dryers because curl is a function of the drying rate from the board surfaces. Board may also differ from paper in relation to M.G. finishes. Pre-dryers and after-dryers are normally used to increase the output determined by the capacity of the M.G. cylinder. The finish required can be obtained by having the board go on to the cylinder at a 60 per cent dryness, and leave 85 per cent dry. The pre-dryers and after-dryers can contribute as follows: from press moisture to cylinder input level and from cylinder exit level to reel dryness. Some of the heavier M.G. paper grades use the same practice. The latter is also used for creped paper towelling where finish is not required.

II.4. Reeling, winding and sheeting

Board machines, other than M.G. machines, usually have two calender stacks in order to obtain the finish required and to control the bulk or thickness of the sheet. It is common, for the first stack, to use "water-doctors", or shallow wooden trays with an open side touching one or more calender rolls, thus adding water to the sheet. The desired board finish can thus be obtained without undue pressure, which would reduce bulk or thickness and consequently stiffness, which is a desirable board quality.

The board machine reel is similar to that of the paper machine. However, small mills seldom use a separate winder because the bulk of the production is sheeted. If reels are needed, they can be slit to width on the machine reel or produced from a very simple and inexpensive rewinder.

Cutter

The cutter is the most important finishing machine. For large capacity board mills, it can be very expensive. It is usually full-width. A double, synchronished fly-knife cutter is used for the production of sheets of standard sizes. Small mills may use a simpler and less expensive technique which involves two operations: cutting followed by guillotining. The added labour cost of guillotining is fairly low, and should be accepted by the discriminating customer.

CHAPTER VI

METHODOLOGICAL FRAMEWORK FOR ESTIMATING
UNIT PRODUCTION COSTS

This chapter is intended to assist persons contemplating the production of pulp and paper by providing them with a methodological framework for the evaluation of alternative small-scale production techniques. The staff of financial institutions, as well as businessmen and government officials, may have their own evaluation methodologies but could still find the following methodological framework useful, especially if they are unfamiliar with pulp and paper production.

I. THE METHODOLOGICAL FRAMEWORK

The methodological framework consists of three parts:

- the determination of the quantities of various inputs used in the process (steps 1 to 5);
- the estimation of the cost of each input and of unit production costs (steps 6 to 13); and
- the profitability of the process considered (steps 14 to 16).

These steps are briefly described below. Producers who wish to identify the most appropriate technique should repeat these steps for each technology which may yield the required output.

Step 1. Determination of the yearly production output. The product may be pulp and/or paper. This figure is a function of such factors as market demand, availability of investment funds and the production technique adopted.

Step 2. Estimation of the quantities of various material inputs for the adopted scale of production. The main materials are:

- straw or bagasse;
- wood pulp;
- various cuttings (cotton, jute);
- caustic soda;
- alum;
- talc;
- starch;
- other liquid or powdered additives (dyes, rosin, china clay);
- fuel; and
- power.

The previous chapters and the examples in this chapter provide information on how these quantities can be estimated.

Step 3. Compilation of a list of required equipment, including spare parts and servicing equipment. The list should include transport equipment, if applicable, as well as testing equipment. Both locally produced and imported equipment should be specified.

Step 4. Estimation of labour requirements. The productivity of the labour force may vary significantly from one country to another. Therefore, the producer should make his own estimate of the appropriate labour requirement. This estimate will depend on the number of shifts per day, working days per week and working weeks per year. Separate estimates should be obtained for skilled and unskilled labour

Step 5. Determination of local infrastructure requirements. This may include:
- land for the plant and buildings;
- land for storage;
- plant building;
- office buildings; and
- other facilities such as power and water.

Step 6 . Estimation of working capital. Apart from purchase of land and equipment, it is necessary to have sufficient initial capital for the purchase of raw materials, payment of wages, and for building up a stock of the product. It is recommended that sufficient working capital be available to cover one month's wages, one month's local raw materials requirements, three months' imported raw materials, one month's fuel and power and allowance for one week's work in progress (value of one week's production).

- 111 -

Step 7. Estimation of the annual depreciation costs of equipment and buildings. These costs depend on the initial purchase price, useful life and prevailing interest rate. Whatever the type of equipment used, it will have a limited life (a recommended life is ten years). The same applies for buildings for which a recommended life is 25 years.

Table VI.2 may be used to estimate these depreciation costs. It provides the discount factors (F) for interest rates up to 40 per cent and expected life periods up to 25 years. Thus, if Z is the purchase price of the equipment (or the cost of a building), the annual depreciation cost is Z/F. For example, if Z = US$1,000, the useful life is ten years and the interest rate 12 per cent, the table provides a value of F equal to 5.65. Therefore, the annual depreciation cost is:

$$Z/F \approx \frac{1.000}{5.65} \approx US\$177.$$

The longer the useful life the lower the annual depreciation costs, and the higher the prevailing interest rate, the higher those costs.

The CIF prices of imported equipment may be obtained from local equipment importers and the cost of buildings and of local equipment may be obtained from local contractors and equipment manufacturers.

Step 8. Estimation of the annual rental rate of land. Land is considered to have an infinite life. Therefore, the annual cost of land, if rented, is its annual rent. If the land is owned by the producer, the annual cost is equivalent to the annual interest on the value of the land.

Step 9. Estimation of the annual cost of materials. The list of materials required for production along with their quantities (from step 2) should now be valued. The cost of the materials can be obtained from local suppliers or, if necessary, from overseas suppliers.

Step 10. Estimation of the total annual labour costs, on the basis of appropriate labour requirements (from step 4) and prevailing local wage rates.

Step 11. Estimation of interest payments on working capital. The working capital required (step 6) involves payment or loss of interest on that sum. Therefore, the annual cost is equal to the interest on the working capital based on the prevailing interest rate.

Step 12. Estimation of total annual production costs. The total annual production cost is equal to the sum of annual depreciation costs for equipment and buildings, annual rental cost for land, annual material costs, annual labour costs and interest payment on working capital. This is the sum of the amounts estimated in steps 7 to 11.

Step 13. Estimation of the unit production cost by dividing the annual production cost by the total number of tonnes of paper produced annually.

Step 14. Estimation of revenues from sales. The total annual sales revenue is estimated by taking into consideration the prevailing local price for the product as well as any sales discounts offered.

Step 15. Estimation of annual net profit before tax. The annual net profit before tax is the difference between annual sales revenue and annual production cost.

II. APPLICATION OF THE METHODOLOGICAL FRAMEWORK

The use of the above methodological framework is illustrated in the following examples.

II.1 Example 1:500 ADT/year paper mill

The framework is used to estimate the cost per air-dry tonne (ADT) of high quality drawing paper. The raw materials include waste cotton cuttings, waste jute cuttings, waste paper and straw or bagasse.

Step 1. Annual production: 500 ADT/year, with the mill operating on a two-shift basis, 300 days/year.

Step 2. Annual requirements of raw materials:
- waste cotton cuttings   250 tonnes
- jute cuttings   150 tonnes
- waste paper   100 tonnes
- straw or bagasse   60 tonnes
- caustic soda   10 tonnes

- bleaching powder	10	tonnes
- rosin	10	tonnes
- china clay	5	tonnes
- titanium oxyde	5	tonnes
- alum	20	tonnes
- starch	5	tonnes
- sizing glue	100	tonnes
- dyes	400	kg

Step 3. Equipment:

Item	Number
- rag chopper	1
- straw chopper	1
- static, atmospheric digesters (oil or gas fired)	2
- beaters	2
- autovat	1
- cylinder mould	1
- drying chamber (gas-fired)	1
- vat for intermittent sheet production	1

Step 4: Labour requirements:

- supervisor	2
- cleaning and cutting personnel	8
- beatermen	6
- cylinder mould operators	12
- pressmen	4
- dryermen	6
- sorting and cleaning personnel	6
- calendering personnel	4
- cutting and packing personnel	4
- indirect labour (manager, accountant, etc.)	9

Step 5: Land and building requirements

- land : 1,400 square metres
- buildings : 300 square metres

Step 6. Working capital requirements: US$

- one month's wages and salaries   2 220
- one month's fuel   750
- one month's supply of local raw materials   16 000
- three month's supply of imported raw materials   7 170
- one week's work in progress   9 760

- total   35 900

Step 7. Annual depreciation costs:

                                                                  US$

- capital costs
        - land   600
        - buildings   26 870
        - equipment   69 350
        - other assets (mostly office equipment)   2 990
        - spares (10 per cent of equipment costs)   7 230

        total   107 040

The discount factors are taken from Table VI.2. For a building life of 25 years, and a 12 per cent interest rate, F = 7.843.

For an equipment life of 10 years and a 12 per cent interest rate, F = 5.65.

The annual depreciation cost of the building is therefore

$$\frac{26,870}{7.843} = US\$3,426.$$

The annual depreciation cost of equipment is

$$\frac{79,570}{5.65} = US\$14,083.$$

The total annual depreciation costs are therefore

$$3,426 + 14,083 = \underline{US\$17,509}.$$

Step 8. Annual rental rate of land:

This is assumed to be 10 per cent of the value of the land and is therefore

$$600 \times \frac{10}{100} = US\$60$$

Step 9. Annual cost of materials

Item	Quantity/year (tonnes)	Unit price (US$)	Cost (US$)
Waste cotton cuttings	250	360	90 000
Jute cuttings	150	297	44 550
Waste paper	100	265	26 500
Straw or bagasse	60	117	7 020
Caustic soda	10	715	7 150
Bleaching powder	10	600	6 000
Rosin	10	1 200	12 000
China clay	5	212	1 060
Titanium oxide	5	1 315	6 575
Alum	20	254	5 080
Starch	5	710	3 550
Sizing glue	5	954	4 770
Dyes	400 kg		16 700
Other chemicals			1 100
Power			3 600
Fuel			9 010
Clothing and maintenance materials			3 715
Total			248 330

Step 10. Annual labour costs

Grade	No. employed	Rate/month (US$)	Cost/annum (US$)
Manager	1	60	720
Clerk	1	48	576
Salesman	1	48	576
Supervisor	2	48	1 152
Cleaning and cutting	8	36	3 456
Beatermen	6	36	2 592
Cylinder mould operators	12	36	5 184
Pressmen	4	36	1 728
Dryermen	6	36	2 592
Sorting and cleaning	6	32	2 304
Calendering	4	36	1 728
Cutting and packing	4	27	1 296
Other indirect labour	6		2 496
Total			26 400

Step 11. Interest payments on working capital.

The interest rate is assumed to be 12 per cent. The working capital from step 6 is US$35,900. Therefore, interest = US$35,900 x 0.12 = US$4,308.

Step 12. Total annual production cost .

The annual production cost is equal to:

US$17,509 + US$60 + US$248,330 + US$26,400 + US$4,308 = US$296,607.

Step 13. Unit production cost.

The unit production cost is equal to

US$296,607 ÷ 500 tonnes = US$593.21 /ADT.

Step 14. Sales revenue.

An average sales price of US$1,015/ADT is estimated. Given the quality of products involved, this is not a high price, especially since the import of equivalent paper would involve transport and duty costs. Therefore, the sales revenue is equal to

US$1,015 x 500 = US$507,500.

Step 15. Net profit before tax:

US$ 507,500 - US$296,607 = US$210,893.

II.2 Example II:25 TPD integrated pulp and paper mill

The major furnish constituents of this mill are straw and bleached softwood pulp for the long fibre support.

Step 1: Annual production : 8,000 ADT/year, of which

- surface-sized paper	3,500 ADT
- lightweight papers	500 ADT
- non-surface-sized writings and printings	4,000 ADT

Step 2. Annual requirements of raw materials.

- straw	17,000 tonnes
- imported pulp	1,400 ADT
- caustic soda	1,480 tonnes
- chlorine	530 tonnes
- lime	215 tonnes
- rosin	400 tonnes
- starch	250 tonnes
- talc	960 tonnes
- alum	400 tonnes
- coal	8,000 tonnes
- power	14.58×10^6 kWh

Step 3. Equipment:

- reconditioned paper machine
- boiler plant
- transformers, motors, etc.
- indigenous equipment for plant
- workshop equipment

Step 4. Labour requirements

- salaried staff, all grades: 116
- skilled wage-earners : 90
- semi-skilled wage-earners : 62
- labourers : 102

Step 5. Land and building requirements:

- land, roads, fencing;
- drains;
- paper machine house;
- pulp mill and boiler house;
- finishing room;
- office block;
- stores;
- workshop; and
- paper stores.

Step 6. Working capital:

	For period of	US$
- salaries	one month	32 000
- utilities (fuel + power)	one month	52 000
- local raw materials	one month	116 000
- imported raw materials	two months	140 000
- work in progress	one week	129 400
	Total	469 400

Step 7. Depreciation costs:

Capital costs	US$
- land	79 050
- buildings	650 150
- equipment	2,946 250
- spares (10 per cent of equipment costs)	294 650
Total	3 970 100

Annual depreciation costs

The discount factors (F) are taken from Table VI.2
-For buildings: life of 25 years at 12 per cent interest:
$$F = 7.843$$
-For equipment: life of 10 years at 12 per cent interest:
$$F = 5.65$$
The annual depreciation cost of the building is therefore:

$$\frac{US\$650,150}{7.843} = US\$82,895$$

The annual depreciation cost of the equipment is:

$$\frac{US\$3,240,900}{5.65} = US\$573,610$$

The total annual depreciation cost is therefore:

US$ 82,895 + US$573,610 = US$656,505.

Step 8. Annual rental rate of land. This is assumed to be 10 per cent of the value of the land and is therefore:

US$79,050 x 0.1 = US$7,905.

Step 9. Annual cost of materials

Item	Quantity/year (tonnes)	Price (US$/tonne)	Annual cost (US$)
Straw	17,000	24	408 000
Imported pulp	1,400 ADT	600	840 000
Caustic soda	1,480	239	353 720
Chlorine	530	78	41 340
Lime	215	36	7 740
Rosin	400	418	167 200
Starch	250	300	75 000
Talc	960	48	46 080
Alum	400	96	38 400
Dyes, misc. chemicals			24 000
Machine clothing			71 600
Maintenance			66 850
Packing and despatch			95 500
Coal	8,000	22	176 000
Power	14.58×10^6 kWh	0.03/kWh	437 400
Water			9 560
Total			2 858 390

Step 10. Annual labour costs: US$384,000

Step 11. Interest payments on working capital. The interest rate is assumed to be 12 per cent. The working capital from step 6 is US$469,400. Therefore, the annual interest payment is

US$469,400 x 12 = US$56,330.

Step 12. Total annual production cost is

US$656,505 + US$7,905 + US$2,858,390 + US$384,000 + US$56,330 = US$3,963,130.

Step 13. Unit production cost (average) is therefore:

US$3,963,130 / 8,000 = US$495.40/ADT

Step 14: Sales revenue

Selling price	Quantity (ADT)	Rate/ADT (US$)	Annual revenue (US$)
Surface-sized paper	3 500	750	2 625 000
Light weight	1 000	150	1 150 000
Ordinary quality	3 500	550	1 925 000
Total sales revenue			5 700 000
Less sales discount (7.5 per cent)			- 427 500
Net sales realisation			5 272 500

Step 15. Net profit before tax:

US$ 5,272,500 - US$3,963,130 = US$1,309,370.

III. COST DATA FOR INDIAN MILLS

The following table (VI.1) has been drawn up from prospectuses issued by three Indian pulp and paper mills, each of 10,000 TPA capacity. This is the maximum capacity for Indian mills if they wish to obtain duty exemption for paper sales and permission to import second-hand machinery.

Each mill uses imported, reconditioned second-hand paper machines which have been officially inspected and guaranteed. Each mill also uses straw, rags and waste paper for the production of furnish. Use of these materials is required by the Government if the mill is to receive government subsidies.

The unit cost of production is very low, but does not include chemical recovery or power generation. It is interesting that, for each mill, compliance with local and national regulations in respect of effluent treatment is claimed. The overall costs in Table VI.1 would probably be 10 per cent higher if new paper machines were used but would still be low by comparison with other projects. Second-hand machines often have features which are not yet available from indigenous suppliers. However, the chief attraction of these machines is that a project can be implemented more quickly.

A proportion of the equity of mills Nos. 2 and 3 has been taken up by the state Development Board. Mills No. 1 and 3 have also obtained subsidies, presumably because they are located in development areas. However, the amount of subsidies, less than 4 per cent of total capital costs, must not have been a major factor in company formation, although they were undoubtedly welcome.

Table VI.1
Cost data from Indian mills

Item	Mill No. 1 Writings & printings	Mill No. 2 Writings & printings	Mill No. 3 Wrapping papers
Capacity	10,000 TPA	10,000 TPA	10,000 TPA
Projected start up	March 1981	February 1981	July 1981
Land purchased	17 hectares	70 hectares	19 hectares
Raw materials	Rice straw (from within a radius of 100 km), waste paper, cotton linter	Wheat/rice straw rags, waste paper	Wheat/rice straw, jute Hessian cuttings, waste paper, rags
Paper machine	Imported, second-hand	Imported, second-hand	Imported, second-hand
Power installed	3,000 KVA, Grid	4,000 KVA, Grid	2,500 KVA, Grid
Water supply	8,200 cu.metres/day	9,250 cu.metres/day	9,090 cu. metres/day
Steam supply	7 T.P.H. 3 coal-fired boilers	10 T.P.H. 2-6 T.P.H. boilers	2-5 T.P.H boilers
Effluent treatment	To Indian standards	To Indian standards	To state requirements
Workforce	100 salaried staff 360 wage earners	500 salary and wage earners	Not stated

PROJECT COST (US$)

Item	Mill No. 1	Mill No. 2	Mill No. 3
Land & site preparation	129 870	177 100	124 580
Buildings	920 900	1 109 800	614 760
Plant & machinery	419 130 (imported) 1 554 900 (indigenous	2 669 410	1 426 570
Engineering	377 800 (inc. duty)	97 400	70 840
Misc. fixed assets	1 617 480	1 284 180	1 108 860
Contingency	354 190	330 600	343 330
Pre-production	646 990	875 100	514 760
Working capital	472 260	392 200	247 350
Total	6 493 520	6 935 690	4 451 050
Annual cost/tonne	649.30	693.50	445.10

FINANCING (US$)

Item	Mill No. 1	Mill No. 2	Mill No. 3
Equity - Promoters	224 320	873 670	
Public	956 320	1 301 510	
Total	1 180 640	2 184 180	1 149 410
Loans	5 076 750	4,427 390 324,120[a]	2 774 540
Subsidy	236 130	-	177 100
Total	6 493 520	6 935 690	4 451 050

[a] Unsecured.

Table VI.2
Discount factor (F)

Year	Interest rate (percentage)																	
	5	6	8	10	12	14	15	16	18	20	22	24	25	26	28	30	35	40
1	0.952	0.943	0.926	0.909	0.893	0.877	0.870	0.862	0.847	0.833	0.820	0.806	0.800	0.794	0.781	0.769	0.741	0.714
2	1.859	1.833	1.783	1.736	1.690	1.647	1.626	1.605	1.566	1.528	1.492	1.457	1.440	1.424	1.392	1.361	1.289	1.224
3	2.723	2.673	2.577	2.487	2.402	2.322	2.283	2.246	2.174	2.106	2.042	1.981	1.952	1.923	1.868	1.816	1.696	1.589
4	3.546	3.465	3.312	3.170	3.037	2.914	2.855	2.798	2.690	2.589	2.494	2.404	2.362	2.320	2.241	2.166	1.997	1.849
5	4.330	4.212	3.993	3.791	3.605	3.433	3.352	3.274	3.127	2.991	2.864	2.745	2.689	2.635	2.532	2.436	2.220	2.035
6	5.076	4.917	4.623	4.355	4.111	3.889	3.784	3.685	3.498	3.326	3.167	3.020	2.951	2.885	2.759	2.643	2.385	2.168
7	5.786	5.582	5.206	4.868	4.564	4.288	4.160	4.039	3.812	3.605	3.416	3.242	3.161	3.083	2.937	2.802	2.508	2.263
8	6.463	6.210	5.747	5.335	4.968	4.639	4.487	4.344	4.078	3.837	3.619	3.421	3.329	3.241	3.076	2.925	2.598	2.331
9	7.108	6.802	6.247	5.759	5.328	4.946	4.772	4.607	4.303	4.031	3.786	3.566	3.463	3.366	3.184	3.019	2.665	2.379
10	7.722	7.360	6.710	6.145	5.650	5.216	5.019	4.833	4.494	4.192	3.923	3.682	3.571	3.465	3.269	3.092	2.715	2.414
11	8.306	7.887	7.139	6.495	5.938	5.453	5.234	5.029	4.656	4.327	4.035	3.776	3.656	3.544	3.335	3.147	2.752	2.438
12	8.863	8.384	7.536	6.814	6.194	5.660	5.421	5.197	4.793	4.439	4.127	3.851	3.725	3.606	3.387	3.190	2.779	2.456
13	9.394	8.853	7.904	7.103	6.424	5.842	5.583	5.342	4.910	4.533	4.203	3.912	3.780	3.656	3.427	3.223	2.799	2.468
14	9.899	9.295	8.244	7.367	6.628	6.002	5.724	5.468	5.008	4.611	4.265	3.962	3.824	3.695	3.459	3.249	2.814	2.477
15	10.380	9.712	8.559	7.606	6.811	6.142	5.847	5.575	5.092	4.675	4.315	4.001	3.859	3.726	3.483	3.268	2.825	2.484
16	10.838	10.106	8.851	7.824	6.974	6.265	5.954	5.669	5.162	4.730	4.357	4.033	3.887	3.751	3.503	3.283	2.834	2.489
17	11.274	10.477	9.122	8.022	7.120	6.373	6.047	5.749	5.222	4.775	4.391	4.059	3.910	3.771	3.518	3.295	2.840	2.492
18	11.690	10.828	9.372	8.201	7.250	6.467	6.128	5.818	5.273	4.812	4.419	4.080	3.928	3.786	3.529	3.304	2.844	2.494
19	12.085	11.158	9.604	8.365	7.366	6.550	6.198	5.877	5.316	4.843	4.442	4.097	3.942	3.799	3.539	3.311	2.848	2.496
20	12.462	11.470	9.818	8.514	7.469	6.623	6.259	5.929	5.353	4.870	4.460	4.110	3.954	3.808	3.546	3.316	2.850	2.497
21	12.821	11.764	10.017	8.649	7.562	6.687	6.312	5.973	5.384	4.891	4.476	4.121	3.963	3.816	3.551	3.320	2.852	2.498
22	13.163	12.042	10.201	8.772	7.645	6.743	6.359	6.011	5.410	4.909	4.488	4.130	3.970	3.822	3.556	3.323	2.853	2.498
23	13.489	12.303	10.371	8.883	7.718	6.792	6.399	6.044	5.432	4.925	4.499	4.137	3.976	3.827	3.559	3.325	2.854	2.499
24	13.799	12.550	10.529	8.985	7.784	6.835	6.434	6.073	5.451	4.937	4.507	4.143	3.981	3.831	3.562	3.327	2.855	2.499
25	14.094	12.783	10.675	9.077	7.843	6.873	6.464	6.097	5.467	4.948	4.514	4.147	3.985	3.834	3.564	3.329	2.856	2.499

A P P E N D I C E S

APPENDIX I

LIST OF EQUIPMENT MANUFACTURERS

The numbers alongside the names and addresses of equipment manufacturers in this appendix correspond to the following equipment:

1. Paper-making machines
2. Board-making machines
3. Pulp chargers
4. Suction rollers
5. Drying cylinders
6. Slitters and cross cutters
7. Winders
8. Roll slitters
9. Complete pulp mills
10. Refiners
11. Agitators and mixers
12. Ancillary equipment
13. Instrumentation and controls
14. Machine clothing (felts, wires, etc.)

CANADA

Albany Engineered Systems Canada, Ltd., Port Alberni, B.C.	1,2,3,4,5
Ashton Press Mfg. Co. Ltd., Montreal, Quebec	6,7,8
Beloit Canada Ltd., Pointe Claire, Quebec	1,2,3,4,5,6,7,9
Black & Clauson-Kennedy Ltd., Owen Sound, Ontario	1,2,3,4,5,6,7,9

Dominion Engineering Works, Ltd.,
Montreal, Quebec   1,2,3,4,5,6,7,

Lightning Greey Mixing Equipment Ltd.,
100 Miranda Avenue,
Toronto I.9   11

McLean H.J.G. Ltd.,
Brantford, Ontario   6,7,8,9,10

Suns Ltd.,
Vancouver, B.C.   1,6,7,8,9

FINLAND
A. Ahlstrom Osakeytio,
S.F. 48601 Karhula   1

Enzo Gutseit Oy,
Helsinki 16   3,9,10

Metex Corporation,
Helsinki 18   1,2,3,4,6,7,8,9,10

Nasin Kiviteollisuus A.I.
Avunien
Tampere 10   4,6,7

Oittivalu Oy,
Oitti   3,9,10

Rauma-Repola Oy,
Helsinki 17   1,2,3,9,10

Tampella, Paper Machinery Division,
P.O. Box 267,
33101 Tampere 10   1

Oy Wartsilla Ab.,
Helsinki 53   1,2,3,4,5,6,7,8,9,10

Yhtym. Paperitehaat Oy,
Jylhavaara,
Valkeatoski   1,2,3,4,5,6,7,8,9,10

FRANCE
Allimand
38140 Rives   1,2,3,4,5,6,8,9,10

Alstom Atlantique
38, av. Kleber,
75784 Paris 16e   1,2,3,4,5,6,

Black Clawson (France),
30, ave. Pierre Curie,
33270 Floirac   1,2,3,4,5,6,7,8,9,10

COFPA,
16002 Angoulême   14

Cognasser,
86, blvd de la République
16000 Angoulême   1,2,3,4,5,6

Escher Wyss France,
208, rue R. Losserand
75014 Paris   1,2,3,4,5,6,7,8,9,10

Hery et Cie,
56, rue de Passy,
75016 Paris   1,2,3,4,5,6

SEIMP,
77, rue de Lourmel,
75015 Paris   1,2,3,4,5,6,7,8,9,10

Sulzer,
Compagnie de Construction mécanique,
Paris   3,10

FEDERAL REPUBLIC OF GERMANY

G. Dorries GmBH, Abt. Papiermaschinen,
D-5160 Duren, Postfach 585   1,2,6

Gustav Reinhard & Co, Maschinenfabrik,
D-5870 Hemer, Postfach 460   1,2,3,4,5

Escher-Wyss GmbH,
D-7980 Ravensburg, Postfach 1380   1

Maschinebau-Werkstatten Niefer GmbH,
D-7532 Niefern-Oschelbronn 1,
Bahnhofstr. 51-53   1,2,4,5

ER-WE-PA Maschinenfabrik und Eisengiesserei
Herbert Karl Schmidt
D-4006 Erkath bei Dusseldorf, Postfach 260   1,2,4,5

Gebr. Bellmer KG Maschinenfabrik,
D-7532 Niefern-Oschelbronn,
Postfach 6   1,2,3,5

Peter Schwabe & Co. K.G. Maschinenfabrik,
D-4130 Moers, Postfach 2267   7,8,

Maschinenfabrik Stahlkontor Weser Lenze K.G.,
D-3250, Hameln 1, Postfach 425   7,8

Carl Krafft & Sohne GmbH,
D-5160 Duren,
Schoellerstrasse 164,   1

Strecker-Bruderhaus,
D-6100 Darmstadt 13, Postfach 130150   1

J.M. Voith GmbH,
D-Heidenheim, Postfach 1940   1

P.J. Wolfe & Sohne GmbH,
D-5160 Duren, Postfach 125

Sandvik Process Systems GmbH,
Salierstrasse 38
D-7012 Fellback bei Stuttgart,
Postfach 4180   12

Wagner GmbH,
D-7410 Reutlingen 1   14

Jagenberg-Werke AG,
D-4000 Dusseldorf, Postfach 1123   6,7,8

INDIA
Eastern Paper Mills Ltd.,
2 Dakshindari Road,
Calcutta 700048   1,9

43.D. Enterprises,
16 Syed Amir Ali Avenue
Calcutta 700017   6,7

Hindon Engineering Works,
Clubley, Bajoria Marg.,
Saharanpur 247001 (U.P.)   1,3

New Allied Capital Industries,
27 Industrial Area,
Chandigarh   2 (up to 10 TPD)

Paper Mill Plant and Machinery
    Manufacturers, Ltd.,
Jogeshwari Estates, 181, S.W. Road,
Bombay 60 NB   1

Saharanpur Engineers and Suppliers,
Ali Mahajan, Khalapur,
Saharanpur   2,3,6,8

Tungbhadra Machinery and Tools Pvt., Ltd.,
43/163 Narasimha Rao Peta,
Kurnool 518004   1,3,6

Indo Berolina Industries Pvt. Ltd.,
5-86 Andheri Kurla Rd.,
Bombay 40018   1

Jessop & Co. Ltd.,
63 Netaj Subhas Rd.,
P.O. Box 108,
Calcutta   1

S.K. Paper Machines Pvt. Ltd.,
Aban House, 5th Floor, Ropewalk St.,
Bombay   1

Utkal Machinery Ltd., Kanshahal 770034
District Sundergarh, Orissa   1

Bakubhai Ambala Pvt. Ltd,
Kaiseri Hind Bldg, 3rd Floor,
Currinbhoy Rd., Ballard Estate,
Bombay 400001   12

Cifoods Ltd.,
Madhupatna,
Cottack, Orissa   12

EIMCO-KCP Ltd.,
2/34 Kodambakham Rd.,
Madras 600034   9,12

Mechano-Industrial Suppliers,
2 Ganesh Chandra Ave.
Calcutta 700013   1,9,12

P.A.S. Engineering Co., Pvt. Ltd.,
8th Floor, Madhuban, 55 Nehru Place,
New Delhi 110019   12

Span Enterprises, Marirawsh Plot No. 115,
Sahakara Nagar No. 2,
Poona 411009   13

Porrits & Spenser (Asia) Ltd.,
308 Kanchenjunga Bldg,
18 Barakhamba Rd.,
New Delhi   14

ITALY

Carcano,
Via Roma 8,
22026 Maslianico   1

Over Meccanica,
Via Toricelli 25,
37100 Verona   1

Recard S.P.A.,
55019 Villa Basilea, Lucca   1

S.N.I.A,
18, via Montebello,
20121 Milano   9

JAPAN

Hitachi Zozen,
6-14 Edobon I-chrome Nishi-Ku,
Osaka 550   9

NETHERLANDS

Berends van Loenen F.A.P.,
Apeldoorn   3,9,10

Engel Machinenfabriek CV. S.W.,
Zaandam   9,10

Kersten NV. Gebr,
Eerbeek   1,4,5,6,7

Robur N.V. Transportwagen,
Honselersdijk   1,4,5,6,7,

Veldhuis & Hulsebos NV.,
Hoogezand   1,3,4,5,6,7,9,10

SPAIN

Sener Technica Industria Y Naval,
Madrid   9

SWEDEN

Alfa-Laval AB,
P.O. Box 500
S-14700 Tumba   1,2,3,4,5,6,7,8,9,10

AB Karlstads Mekaniska Werkstad,
S-65101 Karlstad   1,2,3,4,5,6,7,8,9,10

Mo Do Mekan AB,
P.O. Box 29,
S-891 01 Ornskoldsvik   1,2,3,4,5,6,7,8,9,10

C.J. Wennberg AB,
Ostanvindsgatan 2, Fack
S-651 01 Karlstad 1   1,9

Cellwood Machinery AB,
S-Nassjo   1,10

Elof Hanson,
Goteborg   1,5,7,9,10

AB Hedemora Verskstader,
Hedemora   1,9

Kamyr,
Box 1033,
S-651 15 Karlstad   9

Sunds Defribrator,
P.O. Box,
S-8510 Sundsvall   9,10

Celleco AB Fack,
S-100 52 Stockholm   12

Eurocontrol,
Box 96,
S-66100 Saffle   13

Staafsjo Bruk,
Stavsjobruk,
S-610 23 Kolmarden   12

Lorentzen & Wette AB,
Box 4,
S-163 93 Stockholm   13

Nordiskafelt AB,
Box 161,
301 103 Holmstead   14

Scandiafelt AB,
S-64010 Hogsjo   14

SWITZERLAND

Emile Egger and Co. Ltd.,
2088 Cressier, Neuchatel   12

UNITED KINGDOM

A.B. Graphic Machinery Ltd.,
3, Princess Rd,
Ikley, West Yorks LS29 9NP   4,5,6,7,

Black-Clawson International,
20-26 Wellesley Rd.,
Croydon, Surrey   1

Bielomatik London Ltd.,
Cotswold St.,
London SE27 ODP   1

Dorr-Oliver Co., Ltd.,
Norfolk House, Wellesley Rd.,
Croydon, Surrey   9

Dezurik,
Wrotham Place,
Wrotham, Kent   12

Eldec (Bury) Ltd.,
Brookshaw St.,
Bury, Lancs. BL9 6EF   1,5

Foxboro Yoxall,
Redhill, Surrey RH1 2HL   13

Holder Group,
Brandlesholme House,
Brandlesholme Rd.,
Bury, Lancs. BL8 1JJ   1

Hunt and Moscrop Ltd.,
P.O. Box 8, Apex Works,
Middleton, Manchester M24 IQT   Calenders

C.H. Johnson & Sons Ltd.,
Bradnor Rd.,
Wythenshawe, Manchester M22 4TS   14

J. Kenyon & Son Ltd.,
P.O. Box 35,
Bury, Lancs.   14

Lenox Machine Co., Ltd.,
Mistral No., Parsons La.,
Hickley, Leics. LE10 1XT   1,5

Lippke (U.K.) Ltd.,
11 Alma Rd.,
Windsor, Berks.   13

Masson-Scott Thrissell Eng.,
Easton Rd.,
Bristol BS5 OHE   6,7,8

Vickeries Ltd.,
Norman Rd, Greenwich
London SE09 QJ   12

E.D, Warburton & Co., Ltd.,
Barnbrook Eng. Works,
Bury, Lancs.   1

J. Winterburn Ltd.,
P.O. Box 6, Riverside Works,
Woodhill Rd, Bury, Lancs.   12

UNITED STATES

American Defribrator,
Chrysler Blds,
405 Lexington Avenue,
New York, N.Y. 10017   9

C.E. Bauer Inc.,
Springfield, Ohio 45501   9,10,12

Ingersoll Rand,
150 Burke St,
Nashua, New Hampshire 03061   9,12

Lodding Engineering Ltd.,
P.O. Box 269
Auburn, Mass. 01501   12

Manchester Machine Division
Diamond Int. Corp.,
P.O. Box 509,
Auburn, Mass. 01501   1

Radiclo Noss AB,
2331 San Luis Place,
Greenbay, Wi, 54304   12

Sprout-Waldron Inc.,
Muncey, Pa. 17756   10

Testing Machines Inc.,
400 Bayview Avenue,
Amityville, N.Y. 11701   13

Williams Apparatus Co.,
55 Park Place,
Watertown, N.Y.   13

EUROPEAN LIST OF THE STANDARD QUALITIES OF WASTE PAPER

Group A: Ordinary qualities

Mixed paper and board No. 2 consists of a mixture of various grades of paper and board, without restriction on short fibre content.

Mixed paper and board No. 3 consists of a mixture of the various qualities of paper and board and containing less than 15 per cent short fibre papers, such as newspapers and magazines.

Board cuttings consists of new shavings or cuttings of pressed board or of mixed board free of strawboard.

Mill wrappers consists of packing papers, called reel wrappers such as are used as the outer wrappings for reels, parcels, or reams of new paper, free of bitumen, waxed or plasticised papers.

Corrugated containers consists of used cases or sheets of corrugated board, with or without kraft covers, and a middle of straw or waste paper, free of bitumen, waxed or plasticised papers.

Mixed pamphlets and magazines consists of pamphlets, magazines, catalogues, printed matter and old newspapers, mixed, with or without staples, free of cardboard bound books.

Overissued pamphlets, brochures and magazines in bundles consists of unused pamphlets and magazines, coloured printed matter, free of latex and insoluble glues, in bundles.

Coloured letter (or coloured records) consists of copy and writing paper with or without print in mixed colours.

Light coloured letters (or light coloured records) consists of copy and other writing paper, printed or not, mixed light colours with an allowance of 5 per cent of dark colours. The allowance of papers with a mechanical pulp base shall be agreed between the buyer and the seller.

White pamphlets, without cardboard consists of books or white sheets of paper for printing and may include mechanical pulp, with black print, no colours, free of book covers, linen, synthetic or latex glues.

White pamphlets without cardboard, wood free consists of books or of white sheets of paper for printing, free of mechanical pulp and of coated papers, with black print, no colours, free of book covers, synthetic or latex glues.

Group B - Middle qualities

Old newspapers consists of used newspapers which contain less than 5 per cent of coloured booklets or of publicity pamphlets, free of crumpled paper.

Overissued newspaper in bundles consists of unused daily newspapers printed on white newsprint and which do not contain more than the normal percentage of coloured illustrations, without staples and in original packed bundles.

Overissued newspapers in bales or palletised consists of unused daily newspapers printed on white newsprint which do not contain more than the normal percentage of coloured illustrations, without staples, and in bales or on pallets.

Mixed coloured shavings consists of shavings of magazines or similar printed matter, without restrictions as to colour and of the content of short fibre paper or of coated paper.

Light coloured bookbinders shavings consists of white shavings, without beater-dyed papers printed in mixed colour, made up for the greatest part of chemical pulp with a maximum of 20 per cent of coated paper, unless stipulated otherwise and of a generally bright clear appearance.

Specially light coloured bookbinders shavings consists of white shavings, without beater-dyed paper, printed in mixed colours, made up for the greatest part of mechanical pulp with a maximum of 20 per cent of coated paper unless otherwise stipulated and of a specially bright appearance.

Group C: High qualities

Mixed light coloured shavings (printers shavings) consists of shavings of pale coloured writing and printing paper, made up mostly of white chemical pulp.

Printers shavings of very pale mixed colours consists of mixed pastel shade shavings of writing and printing paper, made up of white chemical pulp, free of dark colours.

Coloured shavings, sorted in colours with mechanical pulp consists of shavings sorted according to shade, bulk coloured, without print, with mechanical pulp.

Coloured shavings, sorted in colours, free of mechanical pulp consists of shavings shorted according to shade, bulk-coloured, without any print and free of mechanical pulp.

Overissued pamphlets, brochures and magazines in bales or palletised consists of unused brochures and magazines printed in colour, free of latex or insoluble glues, in bales or on pallets.

Cuttings of duplex or multi-ply board with a white liner (grey-white) consist of new cuttings or other waste of duplex or multi-ply board, with at least a white liner over a grey interior or back, with or without print.

APPENDIX III

BIBLIOGRAPHY

Ahuja S.P. (ed.): Paper industry in India (New Delhi, Institute of Economic and Market Research, 1977).

Attwood, D.: The search for appropriate technology for the United Kingdom paper and board industry. Paper presented at the UNIDO International Forum on Appropriate Industrial Technology in New Delhi, India, Nov. 1978, UNIDO Doc. No. ID/WG 282/106 (Vienna, 1978).

Chaudhuri, B.P. and Narby, M.: "Operation problems in developing countries", in Pulp and Paper International (Nov. 1979), p. 54.

Commonwealth Secretariat: Paper production report (London, Sep. 1978).

Easton, J.C.: Criteria for selection of small and large-scale pulp and paper mills for the developing countries, small-scale multigrade paper mills and the use of sisal fibres Paper presented at the UNIDO International Experts Group Meeting on Pulp and Paper Technology in Manila, Philippines, Nov. 1980, UNIDO Doc. No. ID/WG/352/31 (Vienna, 1980).

FAO: Advisory Committee on Pulp and Paper: World paper and paperboard consumption outlook (Rome, Sep. 1977).

FAO: Advisory Committee on Pulp and Paper: Report on world pulp and paper capacities (Rome, March 1980).

FAO: Pulp and paper capacities survey-1980-85 (Rome, 1980).

FAO: Guide for planning pulp and paper enterprises, Forestry and Forest Products Studies No. 18 (Rome, 1978).

Greenhalgh, P. and Palmer, E.R.: The production of pulp and paper on a small scale (London, Tropical Development and Research Institute, 1983).

Gordon, I.T. and Currie, D.: "What the small mills need to survive", in Pulp and Paper International (May 1979).

Grant, R.: "Optimum is beautiful small mills can work", in Pulp and Paper International (Dec. 1978).

Hackl, H.J.: "Small mills may be the only solution", in Pulp and Paper International (Dec. 1978).

Hangal Paper Consultants: Report on small-scale pulp and paper industry in India (Rugby, Intermediate Technology Industrial Services, 1979).

National Council of Applied Economic Research: Paper industry, problems and prospects (New Delhi, 1972).

Nylinder, P.: "Wood quality and fibre products" in Appropriate industrial technology for paper products and small pulp mills No. 3, Doc. ID/232/3 (New York, United Nations, 1979).

Reinoso, E; Inglesias, F. and Sanchez, F.: "The use of anthraquinone in the production of bleached kraft", Latin American Congress, Torremolinos, June 1981.

United Nations Industrial Development Organisation: Appropriate Industrial Technology for Paper Products and Small Pulp Mills (New York, United Nations, 1979).

Western, A.W.: Small-Scale Papermaking (London, Intermediate Technology Development Group, 1979).

APPENDIX IV

SELECTED RESEARCH AND DEVELOPMENT INSTITUTIONS

India

- Forest Research Institute,
  Dehra Dun (U.P.)

- Institute of Paper Technology
  Saharanpur (U.P.)

Indonesia

- Celulose Research Institute,
  Bandung

United Kingdom

- Research Association for the Paper and
  Board, Printing and Packing Industries (PIRA),
  LEATHERHEAD, Surrey

Philippines

- Forest Products Research and Industries
  Development Commission,
  Laguna

United States

- Technical Association of the Pulp and Paper
  Industry (TAPPI),
  1 Dunwoody Park,
  ATLANTA, Georgia 30341

GLOSSARY OF TERMS AND ABBREVIATIONS

A.D.   Air-dry, i.e. containing moisture in
                              equilibrium with that of ambient air
                              conditions. An arbitrary definition
                              normally interpreted as 95 per cent dry for
                              paper and 90 per cent dry for market pulp.
                              If dried more, the latter is difficult to
                              re-pulp.

A.D.T.   Air-dry metric tonne.

Bagasse   The fibrous residue from sugar cane after
                              the sugar juices have been extracted by
                              crushing.

B.D.   Bone-dry, i.e. dried completely, sometimes
                              termed oven dry.

B.D.T.   Bone-dry metric tonne.

Beating   The mechanical treatment given by a beater
                              to a suspension of pulp and water to
                              improve the paper-making qualities.

Brown liquor   The term used to describe the final
                              effluent resulting from washing chemically
                              cooked pulp. It contains the combustible
                              lignin and chemical compounds.

Black liquor   Strictly speaking, this term should apply
                              to brown liquor which has been evaporated
                              to the level where it will support
                              combustion, but it is sometimes loosely
                              applied to brown liquor and the evaporated
                              liquor is then called "strong black liquor".

Bleaching	The process of whitening pulp by removing residual organic matter to the desired level.
Blow tank	A vertical tank used to receive cooked pulp under steam pressure from the digester. In the blow tank, the pressure is reduced and the pulp diluted to pumping consistency. Useful heat is also recovered.
Board	The term is used to describe materials with the required stiffness and flatness for making such articles as cartons, boxes, cards. Generally understood to be multi-ply, but can be single-ply, of heavy substance.
Brown stock	Pulp which has been cooked, but not bleached.
Calender	A component of the paper machine used to impart smoothness to the paper or to control thickness.
Chemical pulp	Pulp which has been produced by subjecting the raw fibre source to chemical action under heat, in order to accelerate the process.
Cleaners	Centrifugal separators which remove substances such as sand, grit and small metallic particles which are denser than the rest of the pulp. The separators can be of the low density type (0.25 to 1 per cent), medium density type, more commonly known as liquid cyclones, (1 to 2 per cent) or high density type, (3 to 5 per cent).
Consistency	The term quantifies, by percentage, the amount of fibre in an aqueous slurry. For paper slurries, the expression "density" is also used in the same context.

Consistency regulator	A device used to achieve constant consistency. In practice, it can only reduce consistency by adding water.
Cook	See "chemical pulp".
Couch	This term is used as a noun to describe the last roll on a Fourdrinier before the paper is separated from the wire. Couches can be plain, felt-jacketed or (more commonly today) suction rolls. The expression can also be used as a verb, to describe the action of separating the formed paper sheet from the wire. It can also be used to describe the juxtaposition of a top press roll from the bottom press roll in terms of the horizontal difference in centres between the touching rolls. This definition can also apply to the positioning of the top roll relative to the cylinder mould in a board machine vat.
Crepe	Paper which has been removed from an M.G. cylinder whilst still adhering to it by "doctoring" it off. It is naturally wrinkled or creped and wound in that condition to give elasticity and softness. "Wet" creping can be done off a press roll before drying but is less pronounced.
Chemi-mechanical	The term describes pulp which has been produced by a combination of chemical cooking followed by mechanical attrition. Such pulps have higher yields than purely chemical pulps but cannot be bleached or beaten to the same extent to develop strength. The terms mechano-chemical or semi-chemical are also used to describe these pulps but the latter term has a special connotation for corrugating medium.

Decker	The term describes a machine used to increase the consistency of pulp by removing water. Normally it comprises a cylinder mould in a vat with a top roll from which the thickened pulp is doctored. Deckers may also be "couchless", (i.e. have no top roll).
De-fibring	The process of breaking down market pulp, or recycled paper, into individual fibre.
De-flaker	A unit used to de-fibre a pulp slurry without cutting the fibres and with minimum power input to minimise the effect on drainage qualities.
Digester	The cooking unit for producing chemical or semi-chemical grades of pulp. It may be of the continuous or batch type; the former type can be screw propelled or free flowing while the latter can be static or rotary. Digesters are normally under steam pressure and cook the raw material with the chemicals added.
Doctors	The name is given to blades used for cleaning drying cylinders, wire, or press rolls continuously during operation. Doctors may be simple, stationary or complex, with oscillating mechanism, angle and pressure adjustment, quick-release action and special flexible design to ensure even contact across the full width.
Fabrics	The term is used to describe substitutes for bronze forming-wires, less vulnerable to damage and giving longer service, or for special endless woven units in combination with felts at the presses to improve water removal.

Felts

Press felts are endless woven units used in the press section to protect the paper or board in the tip of the press, to convey it to and from the press and to absorb excess water. On board machines, the long felt is used to pick up successive layers of the board and convey the composite to the presses. The foregoing felts were once all made of wool but today they are more commonly made from synthetic materials. Dryer felts are not endless but seamed in position; they serve to hold the paper or board firmly against the dryer surface. They were originally of cotton asbestos material but have today been virtually superseded by "screens" (heavy plastic woven material) which last longer and improve the drying rate.

Fillers

These are materials used to improve the opacity, brightness and printing qualities of paper. China clay, talc, and for expensive papers titanium oxide are the most common fillers.

Flow-box

Prepared stock, diluted to the correct consistency, is introduced to the paper-machine wire through a flow-box, the function of which is to keep the fibres in suspension and present them evenly through a "slice" to the wire.

Foils

Specially profiled stationary plastic blades used to support the upper horizontal section of a Fourdrinier wire and to promote drainage. Their action generates less vacuum than rolls and formation can be improved.

Formers

Machines differing from the fundamental Fourdrinier or vat for the forming of paper or board. There are several proprietary units which may be divided into two classes: twin wire formers on which the paper is formed at high speed between two synchronised wires, or cylindrical units in which a flow-box is used to deposit the pulp slurry on to a wire-wound cylinder. The latter category ranges from simple, vacuum less units to highly sophisticated units with internal vacuum boxes and additional wires.

Fourdrinier

The original continuous paper forming unit, comprising an endless mesh of bronze wire or plastic running over rollers. This forming machine is still the most commonly used and versatile in performance.

Furnish

Blend of fibres and additives which makes up the final paper sheet.

g/m^2

Grams per square metre: a measure used to define paper or board substance as weight per unit of area.

Green liquor

This expression is used to describe the solution of the smelt obtained from burning evaporated black liquor. After clarification, green liquor can be reconverted to cooking liquor by re-causticising.

Groundwood

Pulp obtained by subjecting wood to grindstones in the presence of water. Also "mechanical pulp". The yield is high but the strengh is low and cannot be much improved by refining. Groundwood is used for the cheaper, less permanent types of paper such as newsprint. "Refiner groundwood" is obtained by using refiners instead of grindstones and the resultant pulp is stronger. "Thermo-mechanical" groundwood uses refiners but also the heat generated to produce a high yield pulp with a strength superior to that of groundwood or refiner groundwood.

Hardwood

Wood from deciduous trees, normally short-fibred.

Hydrapulper

A machine used to slush waste paper or imported pulp. It is a proprietary trade name for one particular design. However, it is now often used to describe units with similar functions.

Kraft

The term describes pulp produced by the sulphate process, or paper made from such pulp. The paper is stronger than that produced by most other processes.

Mechanical pulp

Groundwood

M.G.

Machine-glazed. Under the machine glazing technique paper is dried by applying one side to a large cylinder under pressure. When sufficiently dry, it leaves the cylinder and has a finish equivalent to that of the cylinder surface, which is ground. A machine-glazing cylinder is more commonly known as a "Yankee" cylinder.

Pick-up	Action of lifting the wet sheet from a Fourdrinier or vat full-width onto a felt which conveys it to the presses. Vacuum pick-ups use a suction roll inside the felt.
Potcher	Shallow tub with a mid-feather used for washing rag-based pulp.
Refiners	Machines which subject a pulp slurry to mechanical treatment in order to develop characteristics appropriate to the end-product.
Size	Liquid added to the paper to control ink penetration into the final product. Size may be an emulsion of rosin added to the stock, a starch solution added to the paper surface or gelatine added to the surface.
Softwood	Wood from coniferous trees, normally having fibres that are long by comparison with those of hardwood.
Suction boxes	Wire (q.v.) suction boxes are of rectangular cross-section and full machine width and have a slotted or perforated wooden or synthetic top supporting the wire. The rear end of the box is connected to a vacuum source to promote drainage. Suction boxes are also used to remove water from press felts.
Suction roll	Hollow metal roll perforated with a close pattern of drilled holes and having one or more vacuum compartments inside. Used for the final stage of water removal from a Fourdrinier, or rubber-covered as a press roll .

Supercalender	Machine used in a secondary process to impart higher than normal finish to paper or board. It is a vertical stack of rolls, made alternately of chilled iron and compressed fabric.
Sulphite pulp	Sulphite pulp is pulp that is chemically cooked by the acid, sulphite process. It was once universal but has been replaced for most purposes by kraft pulp which permits chemical recovery and is stronger.
Winder	The machine following a paper or board machine. Its purpose is to rewind machine reels to reels of narrower width and smaller diameter and to remove faulty material.
Wire	Full width, endless wire mesh belt on which the paper is first formed. Originally, wires were all made of bronze, hence the name. Synthetic plastic filament wires are increasing in popularity.
Vat	The trough in which a cylinder mould runs in a board machine is, strictly speaking, the vat. The term is also loosely used to describe the complete unit, including the vat, the cylinder mould and the couch roll.

www.ingramcontent.com/pod-product-compliance
Ingram Content Group UK Ltd.
Pitfield, Milton Keynes, MK11 3LW, UK
UKHW050057040626
6223IPUK00005B/100